T0245153

Applied ethology 2013:

Understanding behaviour to improve livelihood

ISAE2013

Proceedings of the 47th Congress of the International Society for Applied Ethology

2 – 6 June 2013, Florianopolis, Brazil

Understanding behaviour to improve livelihood

edited by:

Maria José Hötzel

Luiz Carlos Pinheiro Machado Filho

Wageningen Academic
P u b l i s h e r s

ISBN: 978-90-8686-225-2
e-ISBN: 978-90-8686-779-0
DOI: 10.3921/978-90-8686-779-0

First published, 2013

© Wageningen Academic Publishers
The Netherlands, 2013

Foreword

It's a great honour to host the Congress of ISAE once again. In the year 2000 we had the challenge to host the 34th ISAE's Congress, which was held for the first time outside Europe and North America. Having its congresses in different regions of the world, ISAE helps applied ethology – and animal welfare – to develop even further. It also has a positive impact on ISAE, helping it become truly international.

ISAE's Congress is a unique moment where scientists, University faculty, students and professionals present their work, exchange their views, teach and learn new ideas, concepts and information. In the 47th Congress of the International Society for Applied Ethology, there were 177 papers submitted and 140 approved to be presented as plenary (6), oral (71) and poster (63) presentations. We are proud to have a Brazilian give the Wood Gush Memorial lecture this year. Dr. Paulo S. Oliveira is a well-known professor from the University of Campinas, and will be presenting his views on the behavioural ecology of insects and ant-plant-herbivore interactions. A fundamental theme – behavioural ecology – to understand Applied Ethology. We also make a tribute to the late Professor César Ades, who was probably the most important Brazilian ethologist of all time. He did some fantastic scientific work, and was also the main person developing Ethology in Brazil, always in an enthusiastic way.

The theme of the 47th congress is *Applied Ethology: understanding behaviour to improve livelihood*. We chose this theme because we believe that Applied Ethology is in the foundations of Animal Welfare, an issue of paramount importance. It is not only an ethical concern, but it is also about the quality of everyday life of billions of animals. To really make a difference in the quality of life of animals, we have to go beyond public satisfaction or marketing strategies. Scientific information can provide solid grounds for decision-making on animal welfare and related issues concerned to quality of life.

This Congress has been organised on a volunteer basis only. We acknowledge first people that volunteered to organise the Congress – in the organising, scientific and ethical committee, the reviewers, and specially Maria C. Yunes. We also acknowledge our sponsors FAPESC (Fundação de Amparo à Pesquisa e Inovação do Estado de Santa Catarina), CAPES (Coordenação de Aperfeiçoamento de Pessoal de Nível Superior) and CNPq (Conselho Nacional de Desenvolvimento Científico e Tecnológico), and the attendance awards from the Humane Society International Award, Nestlé Purina Award and ISAE for the Congress Attendance Fund.

Why did we accept the challenge and the work of organising the ISAE's Congress again? Because we love ISAE's congresses and all the interesting people attending them. We hope you enjoy the Congress and have a lovely time in Florianópolis.

Luiz Carlos Pinheiro Machado Filho and Maria José Hötzel

Acknowledgements

Congress Organising Committee

Luiz Carlos Pinheiro Machado Filho (*Chair*)
Maria José Hötzel
Maria Cristina Yunes
Daniele Cristina da Silva Kazama
Ricardo Kazama

Scientific Committee

Maria José Hötzel (*Chair*)
Luiz Carlos Pinheiro Machado Filho
Rosangela Polleto
Selene Nogueira

Ethics Committee

Ian Duncan (*Chair*)
Donald Lay Jr.
Kristin Hagen
Anna Olsson
Maria José Hötzel
Alexandra Whittaker
Francois Martin

Support Committee

Aurilédia Batista Teixeira, Bruna Raizer, Cibele Longo, Clarissa Cardoso, Denise Leme, Fabiellen Cristina Pereira, Fernanda Katharine de Souza Lins Borba, Jéssica Rocha Medeiros, Patrizia Ana Bricarello.

Design

Congress website: Paula Pinheiro Machado
Proceedings cover: Paula Pinheiro Machado

Referees

Michael Appleby
Gregory Archer
Harry Blokhuis
Xavier Boivin
Sylvie Cloutier
Mike Cockram
Anne-Marie de Passillé
Giuseppe De Rosa
Trevor Devries
Ian Duncan
Catherine Dwyer
Monica Elmore
Hans Erhard
Carla Forte Maiolino Molento
Brianna Gaskill
Martina Gerken
Jessica Gimpel
Anke Gutmann
Lorenz Gygax
Laura Hänninen
Marjorie Haskell
Edna Hillmann
Luciana Honorato
Maria José Hötzel
Margit Bak Jensen
Larry Katz
Linda Keeling
Nina Keil
Paul Koene
Jan Langbein
Donald Lay Jr.
Denise Pereira Leme
Yuzhi Li
Lena Lidfors
Cibele Longo

Xavier Manteca
Jeremy Marchant-Forde
Lindsay Matthews
Marie-Christine Meunier-Salaün
Lene Munksgaard
Ruth Newberry
Keelin O'driscoll
Anna Olsson
Edmond Pajor
Mateus Paranhos da Costa
Carol Petherick
Luiz Carlos Pinheiro Machado Filho
Anthony Podberscek
Rosangela Poletto
Jean-Loup Rault
Bas Rodenburg
Jeffrey Rushen
Heike Schulze Westerath
Janice Siegford
Selene Siqueira da Cunha Nogueira
Hans Spoolder
Carolyn Stull
Stephanie Torrey
Cassandra Tucker
Rodolfo Ungerfeld
Anna Valros
Elsa Vasseur
Isabelle Veissier
Kathalijne Visser
Eberhard von Borell
Marina von Keyserlingk
Susanne Waiblinger
Daniel Weary
Françoise Wemelsfelder
Christoph Winckler

Sponsors:

UNIVERSIDADE FEDERAL
DE SANTA CATARINA

FAPESC
Fundação de Amparo à Pesquisa e
Inovação do Estado de Santa Catarina

CAPES

*Conselho Nacional de Desenvolvimento
Científico e Tecnológico*

Students Awards

Celebrating Animals | Confronting Cruelty
Worldwide

HUMANE SOCIETY
INTERNATIONAL

Program at a glance

Sun 2nd June	Centro de Eventos da UFSC	
3.00	Registration Open	
6.00	Opening Ceremony	
6.45	David Wood-Gush Memorial Lecture	
7.50	Welcome reception	

Mon 3rd June	Costão do Santinho Resort	
	Theater 1	Theater 2
8.30	Plenary 1	
9:10	1 minute Poster Presentations	
9.50	Coffee Break & Posters	
10.30	Session 1: Housing and behavior	Session 2: Zoo and wild animals
	Session 3: Free papers	Session 4: Temperament and behaviour
12.15	Lunch	
2.00	César Ades' Tribute	
2.30	1 minute Poster Presentations	
2.50	Posters	
3.30	Coffee Break	
4.00 – 6.00	Session 5: Pig behaviour and welfare	Session 6: Preferences and motivation
	Session 7: Pig behaviour and welfare	Session 8: Dairy calf behaviour

Tues 4th June	Costão do Santinho Resort	
	Theater 1	Theater 2 `
8.30	Plenary 2	
9.10	1 minute Poster Presentations	
9.50	Coffee Break & Posters	
11.00	Session 9: Companion animals	Session 10: Social Behaviour
12.30	Lunch	
2.30	Plenary 3	
3.10	Session 11: Cognition and emotion	
3.55	Coffee Break & Posters	
4.30 – 6.00	AGM of the ISAE	
8.30	Congress Dinner	

Wed 5th June	Costão do Santinho Resort	
	Theater 1	Theater 2
9.00	Excursions	

Thurs 6th June	Costão do Santinho Resort	
	Theater 1	Theater 2
8.30	Session 12: Sheep Behaviour	Session 13: Free Papers
10.15	Coffee Break & Posters	
10.45	Plenary 4	
11.25	Session 14: Laboratory Animals	
12.30	Lunch	
2.30	Plenary 5	
3.10	Session 15: Euthanasia	
4.10	Coffee Break & Posters	
4.40	Session 16: Dairy cows' use of outdoor environments	
5.40	Closing of Conference	
	Farewell Party	

Scientific Program

Sunday 2nd June – Centro de Eventos da UFSC

3.00	Registration Open
6.00	Opening Ceremony

6.45 **David Wood-Gush Memorial Lecture**

Behavioral ecology at the ant-plant-herbivore interface: Ants mediate colonization decisions in tropical butterflies

Paulo S. Oliveira

Chair: Luiz Carlos Pinheiro Machado Filho

7.50	Welcome Reception

Monday 3rd June – Costão do Santinho Resort

8.30 **Plenary 1:**
A new science of animal suffering
Daniel M Weary
Chair: Maria José Hötzel

9.10 1 minute Poster Presentations

9.50 **Coffee Break & Posters**

10.30 **Parallel Sessions**

	Session 1: Housing and behaviour *Chair*: Rebecca Doyle	Session 2: Zoo and wild animals *Chair*: Lindsay Matthews
10.30	Does rubber flooring influence postural and spatial behaviour in gestating sows? **Julia Calderón Díaz**	Pekin duck social behavior: time budgets, synchrony and preferential associations **Maja Makagon**
10.45	How much straw do slaughter pigs need to meet the need for exploration? **Lene Jull Pedersen**	The temperament of peccaries (Mammalia, Tayassuidae) applied to their management in captivity **Selene Nogueira**
11.00	The use of perches by broiler chickens **Paula Baker**	Behavioral response of guanacos (Lama guanicoe) to tourists in Torres del Paine National Park, Chile **Beatriz Zapata**
11.15	The impact of space availability in movement patterns of dairy sheep during pregnancy **Xavier Averós**	A multi-zoo study on the effect of manipulated visitor behaviour on zoo meerkats **Sally Sherwen**

Monday 3rd June – Costão do Santinho Resort

	Session 3: Free papers *Chair:* Rebecca Doyle	Session 4: Temperament and behaviour *Chair:* Lindsay Matthews
11.30	Evaluation of rubber rings coated with lignocaine for pain relief during castration of lambs **Mairi Stewart**	Temperamental turkeys: do tonic immobility, open field and novel object test responses change? **Marisa Erasmus**
11.45	Achieving a high quality of life for farm animals requires provision, rather than deprivation **Rebecca Sommerville**	Temperament traits of sows in two group housing systems **Yolande Seddon**
12.00	Piagetian object permanence abilities in goat kids **Rachel Chojnacki**	Identification of temperamental categories in horses **Malin Axel-Nilsson**
12.15	Lunch	
	Tribute to César Ades	
	Selene Nogueira	
	Chair: Angélica Da Silva Vasconcellos	
2.30	1 minute Poster Presentations	
2.50	*Poster Session*	
3.30	**Coffee Break**	
4.00	**Parallel Sessions**	
	Session 5: Dairy calf behaviour *Chair:* Jeffrey Rushen	Session 6: Dairy calf behaviour *Chair:* Jeffrey Rushen
4.00	The evaluation of behavioural and clinical parameters in assessing weaner pig recovery on farm **Rebecca Wilson**	The effect of level of social contact in dairy calves on behavior and health **Margit Bak Jensen**
4.15	Differences between female and castrated male pigs in their response to novelty **Inonge Reimert**	Rearing substrate affects play behaviour in an arena test in dairy calves **Mhairi Sutherland**
4.30	Pigs make up their mind early in life: behaviour in novelty tests and blood serotonin **Winanda Ursinus**	Time spent eating solid feed predicts intake in milk fed dairy calves **João H. C. Costa**
4.45	Effects of alginate and resistant starch on feeding patterns, behavior and growth in growing pigs **Carol Souza da Silva**	The effect of competition for teat access on feeding patterns of dairy calves **Emily Miller-Cushon**

Monday 3rd June – Costão do Santinho Resort

	Session 7: Pig behaviour and welfare *Chair*: Jeremy Marchant-Forde	Session 8: Preferences and motivation *Chair*:Jeffrey Rushen
5.00	The effect of different nutritional regimes on locomotory ability and lying behaviour of replacement **Amy Quinn**	Chopped or long roughage: what do calves prefer? **Laura Webb**
5.15	Group-housed sows who engage in aggression after mixing have reduced injuries and gain more weight **Megan Verdon**	Dairy cattle preferences for sprinklers delivering different water volumes **Jennifer Chen**
5.30	The effects of non-uniform environmental conditions on piglet mortality and behavior of sows **Gabriela Morello**	Housing type affects the lying behaviour of dairy cows given access to an outside 'loafing paddock' **Fritha Lagford**
5.45	Indirect genetic effects and housing conditions influence aggressive behaviour in pigs **Irene Camerlink**	Assessing motivation to access a food reward in laying hens while kept in a social group' **Carlos E. Hernandez**

Tuesday 4th June

8.30	Plenary 2:
	Many mammal species are unsuitable for companion animal
	Paul Koene
	Chair: Jessica Gimpel Rivera
9.10	1 minute Poster Presentations
9.50	**Coffee Break & Posters**

11.00	**Parallel Sessions**	
	Session 9:	*Session 10:*
	Companion Animals	Social Behaviour
	Chair: Jessica Gimpel Rivera	*Chair:* Marina Von Keyserlingk
11.00	Emotional and behavioural responses in gun dogs during blind-retrieving tasks in a T-maze **Donald Broom**	Can pigmentation and hair cortisol predict social strategies in goats? **Judit Vas**
11.15	Dog owners' perception of their dogs' responses to loss of a canine companion **Leticia Fanucchi**	Reducing aggressive behaviour in young piglets by cognitive environmental enrichment **Lilia Thays Sonoda**
11.30	Welfare of community dogs in Campo Largo-PR/ Brazil: preliminary results **Larissa Rüncos**	Inter-logger variation of spatial proximity devices: consequences for animal social networks **Natasha Boyland**
11.45	Do cats' initial responses to a confinement housing environment persist upon re-exposure? **Judith Stella**	Association of social rank with health, reproduction, and milk production of dairy cows **Marcia Endres**
12.00	Is the strange situation procedure a reliable method to use when investigating attachment in dogs? **Therese Rehn**	Comprehension of human social cues in young domestic pigs **Christian Nawroth**
12.15	Development and validation of a behavioural test to discriminate between bold and shy cats **Jacklyn Ellis**	How do pigs vocally communicate: a graded or continuous system? **Céline Tallet**
12.30	Lunch	

Tuesday 4th June

2.30	**Plenary 3:** **Protective mother hens:** **Cognitive and emotional influences on the avian maternal response**
	Joanne Edgar
	Chair: Anne Marie de Passillé
3.10	**Oral Sessions**
	Session 11:
	Cognition and Emotion
	Chair: Anne Marie de Passille
3.10	Dairy calves show negative judgement bias following hot-iron disbudding
	Heather Neave
3.25	Individual housing impairs reversal learning in dairy calves
	Ruã Daros
3.40	Developing a titration method to define individual decision difficulty in laying hens
	Anna Davies
3.55	**Coffee Break & Posters**
4.30-6.00	AGM of the ISAE
8.30	**Congress Dinner**

Wednesday 5th June

9.00	**Excursions**

Thursday 6th June

8.30	Parallel Sessions	
	Session 12: Sheep Behaviour *Chair:* Agustín Orihuela	**Session 13:** Free Papers *Chair:* Celine Tallet
8.30	The behaviour during the appetitive motivational state for a reward differ depending on the reward **Claes Anderson**	Attack intensity of pest flies and behavioural responses of pastured dairy cows **Carrie Woolley**
8.45	Qualitative and quantitative behavioural assessment in sheep during feeding motivation tests **Lindsay Matthews**	Behaviour and health of different turkey genotypes with outdoor access **Jutta Berk**
9.00	Testing for attention biases in sheep **Rebecca Doyle**	The body-behavior connection: associations between WQ® measurements and non-cage laying hen behavior **Courtney L Daigle**
9.15	Ewes direct more maternal attention toward lambs that express the most severe pain-related responses **Katarzyna Maslowska**	On Farm Assessment of Milking Behavior in Dairies with Automatic Milking Systems **Janice M. Siegford**
9.30	Stress during pregnancy affect maternal behaviour specifically in passive stress-responsive ewes **Marjorie Coulon**	Preliminary evidence of an altered serotonin metabolism in the prefrontal cortex of tail biting pigs **Anna Valros**
9.45	Neuro-behavioral reactions to physical stimuli varying in valence in sheep of different mood states **Sabine Vögeli**	The effect of grass white clover and grass only swards on dairy cows grazing behaviour and rumen pH **Daniel Enríquez**
10.00	Does handling experience alter the response of sheep to the presence of an unfamiliar human? **Susan Richmond**	Changes in exploratory feeding behaviour as an early indicator of metritis in dairy cattle **Juliana M. Huzzey**
10.15	**Coffee Break & Posters**	

10.45	**Plenary 4:**
	Nest building as an indicator of illness in laboratory mice

Brianna N. Gaskill

Chair: Anna Valros

11.25	**Oral Sessions**
	Session 14:
	Laboratory Animals
	Chair: Anna Valros

11.25	Predicting, preventing, and treating barbering behavior in C57BL/6 mice
	Giovana Vieira
11.40	Evaluation of the well being of rats housed in multilevel caging with red tinted polysulfone
	Melissa Swan
11.55	Comparing daylight behavioural time budgets of young and senescent laboratory mice
	Jessica Gimpel
12.10	Intra- and inter-test consistency of fear tests for adult rabbits
	Stephanie Buijs
12.30	**Lunch**

2.30	**Plenary 5:**
	Effects of depression score on welfare implications of CO_2 and argon gas euthanasia of piglets

Larry Sadler

Chair: Donald Broom

3.10	**Oral Sessions**
	Session 15:
	Euthanasia
	Chair: Donald Broom

3.10	Assessing the humaneness of mechanical methods for killing poultry
	Jessica E. Hopkins
3.25	Zebrafish aversion to chemicals used as euthanasia agents
	Devina Wong
3.40	A search for humane gas alternatives to carbon dioxide for euthanizing piglets: A piglet perspective
	Donald C. Lay Jr.
3.55	Evaluation of microwave application as a humane stunning technique based on electroencephalography
	Jean-Loup Rault
4.10	**Coffee Break & Posters**
4.40	**Oral Sessions**

Thursday 6th June

	Session 16: Dairy cows' use of outdoor environments *Chair:* Luiz Carlos Pinheiro Machado Filho
4.40	The effect of being outdoors during the day or night on walking and lying behaviour in dairy cows **Mark S. Rutter**
4.55	Dairy cow preference for open air exercise during winter under Eastern Canada climatic conditions **Elsa Vasseur**
5.10	Dairy heifer preference for being indoors or at pasture is affected by previous experience **Priya R. Motupalli**
5.25	Motivation for Access to Pasture in Dairy Cows **Andressa A. Cestari**
5.40	Closing of Conference
	Farewell Party

Table of contents

Session 02. Zoo and wild animals

Session 03. Free papers

Session 04. Temperament and behaviour

Session 05. Pig behaviour and welfare

Session 06. Dairy calf behaviour

Session 07. Pig behaviour and welfare

Session 08. Preference and motivation

Session 11. Cognition and emotion

Session 12. Laboratory animals

Session 13. Sheep behaviour

Session 14. Free papers

Session 15. Euthanasia

Session 16. Dairy cows' use of outdoor environments

Poster session

Behavioral ecology at the ant-plant-herbivore interface

Paulo S. Oliveira

Instituto de Biologia, Departamento de Biologia Animal, Universidade Estadual de Campinas, 13083-862 Campinas SP, Brazil; pso@unicamp.br

Foliage represents a major zone of biological interaction in terrestrial ecosystems, and herbivores have to cope with the frequent threat of predation. In insect herbivores, the capacity to make appropriate decisions during host plant selection is a crucial behavioral trait. Because natural enemies are abundant on foliage, herbivores face a major dilemma: they need to find a safe space for their offspring. Information about predation risks is critical for egg-laying females of insect herbivores, and natural selection may favor the ability to detect predators and to select enemy-free foliage when offspring mortality risk is high. Ants are the most important predators on tropical plants and many aggressive foliage-dwelling species actively hunt on insect herbivores. Therefore, by detecting predacious ants prior to plant colonization, gravid females should be able to shift egg-laying to less risky foliage. Some types of insect herbivores, however, not only circumvent ant predation but even attract ants for their own benefit: they secrete liquid rewards (honeydew) and attract tending ants that behave aggressively as bodyguards. For such myrmecophilous herbivores (literally 'ant-lovers'), natural selection on gravid females may favor the ability to detect ant mutualists prior to egg-laying so as to select ant-occupied plant locations that would improve offspring survival. Our research group investigated colonization decisions by two butterfly species whose offspring (caterpillars) experience two contrasting ecological scenarios in the Brazilian savanna: antagonism with predatory ants, and mutualism with honeydew-gathering ants. Field experiments show that detection of ants by gravid females to the benefit of larval offspring – either through avoidance of ant predation, or through protection via ant mutualists – may have represented an important evolutionary step in the process of host plant selection in butterflies living in ant-rich environments. While large herbivores such as gazelles and zebras are known to distinguish by sight among lions, cheetahs, and hyenas in the African savanna (and react differently to each), we are only beginning to assess the cues eliciting behavioral decisions by insect herbivores in the context of their interspecific interactions (supported by FAPESP, CNPq).

A new science of animal suffering

Daniel M. Weary
University of British Columbia, Animal Welfare Program, 2357 Main Mall, Vancouver, BC, V6T 1Z4, Canada; danweary@mail.ubc.ca

In much of the recent animal welfare literature the word suffering is used simply as an adjunct (as in 'pain and suffering') or to emphasize that the animal consciously perceives pain or some other negative affect. A stronger usage of the term implies that the negative feelings are prolonged, high intensity or both, but without any clear line to distinguish when suffering begins. Researchers in human medicine have developed more explicit definitions of suffering that also reference concurrent negative feelings (including fear, anxiety, sadness and depression) and the patient's ability to cope. Applying this broader definition of suffering to animal welfare will require a new approach to the research we do. I outline the basis for a new science of animal suffering; this research will require not only the assessment of negative affective states but also an assessment of how concurrent negative states interact, a general assessment of the animal's emotional health and its ability to cope with adversity.

Many mammal species are unsuitable for companion animal

Paul Koene, Bert Ipema and Rudi De Mol
Wageningen University and Research center, Livestock Research, De Elst 1, 6708 WD Wageningen, the Netherlands; paul.koene@wur.nl

Between 2009 and 2013, a framework has been developed to enhance the transparency and objectivity of detecting the suitability of mammal species as companion animals. The framework is based on citations from literature in a database. The citations have been related to species characteristics summarized in 12 criteria (space-, time-, metabolism-, shelter-, sex-, comfort-, biosocial, information-related and other behaviour, welfare, health and human-animal interaction) with 83 subcriteria. Data about the species in the wild and about the species in captivity have been classified separately in the database. An internet inquiry for animal owners was used to collect bibliographical and numerical information about the species kept as companion animal in the Netherlands. Ninety mammal species are being kept. For each species the literature was searched in a standardized way using web of science, encyclopaedia's, internet sources, specialist literature and information provided by stakeholders. Based on the literature citations the strength of needs or risks of the mammal species was valued by a group of eight scientists from the fields of zoology, ethology, ecology and applied animal welfare science. Their combined judgements were presented in graphs to scientists and stakeholder representatives, of whom the assessment of the species' suitability was recorded. In addition, the group of eight scientist also estimated the presence or absence of favorable and unfavorable characteristics that may exapt a species to adapt readily to a human environment. On base of the favorable and unfavorable characteristics the so-called Pet Suitability Index (PSI) was calculated. The PSI indicated for each species its adaptability to the human environment ranging between were 0.249 and 0.738. Cat species (e.g. the leopard cat en the serval) had a low PSI (0.325, 0.348) and cavies and camels (e.g. capybara and Brazilian cavy, lama and Bactrian camel had a high PSI (0.704, 0.738, 0.702, 0.651). The final judgements of the needs, limitations, health, welfare and environmental risks of the mammal species investigated indicated that few (n=6) mammal species appear to be suitable to be kept without specialist knowledge and skills, many species are unsuitable to be kept (n=61) and for some species the outcomes are not yet decisive (n=23). Of many species crucial information about characteristics was lacking. Many mammal species showed a positive relation between estimated PSI and estimated actual suitability, but also exceptions were found. The dynamical framework summarizes knowledge of mammal species, assesses the suitability for companion animal, stimulates discussions about suitability of keeping mammals as pets and species-specific research in nature and captivity.

Protective mother hens: cognitive and emotional influences on the avian maternal response

Joanne Edgar, Elizabeth Paul and Christine Nicol
University of Bristol, Clinical Veterinary Science, Langford House, Langford, BS40 5DU, United Kingdom; j.edgar@bristol.ac.uk

Domestic animals may be frequently exposed to situations in which they witness the distress or pain of conspecifics and the extent to which they are affected by this will depend on their capacity for empathy. Emotional empathy occurs when one individual (the observer) detects the emotional responses of another individual (the demonstrator), in response to a stimulus, triggering a matching emotional response in the observer. Our previous research showed that domestic hens (*Gallus gallus domesticus*) show behavioural and physiological responses when witnessing their chicks in a state of mild distress, prompting on-going research to assess whether these are associated with a valenced, emotional response. Another fundamental question is whether a mother hen's response to chick distress is mediated by the hen's knowledge about the situation, or more simply, by distress cues from her chicks. We therefore investigated how manipulating hen and chick knowledge influences hens' responses to chick distress. Each hen's brood of chicks was split into three groups, based on whether they had the same, opposite or no knowledge about a potentially threatening situation (environmental cues signalling air puff administration). We compared hen behaviour, vocalisations and physiological responses to actual and perceived threat to their chicks. Hens responded behaviourally when they perceived their chicks to be threatened, regardless of the chicks' reactions to the situation. Chick behaviour was influenced by the hens' perceptions, with all chick groups responding when in the environment that the hen associated with threat. We suggest, therefore, that the protective maternal response of domestic hens is not solely driven by chick distress cues; rather, hens integrate these with their own knowledge to produce a potentially adaptive, flexible and context-dependent response. In this session we discuss the concepts and methodology underpinning our recent studies on the behavioural, physiological and valenced responses of domestic hens to mild chick distress, and suggest direction for future research to determine the extent to which animals show the capacity for empathy.

Nest building as an indicator of illness in laboratory mice

Brianna N. Gaskill and Kathleen Pritchett-Corning
Charles River, Research Models and Services, 251 Ballardvale St, Wilmington, MA 01887, USA;
brianna.gaskill@crl.com

Laboratory housed mice provided crinkled paper nesting material build fully enclosed nests, reducing cold stress, but preventing direct daily observations by animal care staff. It has been suggested that ill mice are not allowed within the nest by conspecifics and do not engage in nest building activity. We hypothesized that nest shape and the exclusion of ill mice during inactivity could be used as tools to identify ill mice. Forty two female C57BL/6NCrl mice were provided 10 g of nesting material and assigned to a social treatment (group of 3 or solitary; n=3). Food and water was provided *ad libitum*. Lipopolysaccharide (LPS) IP injection was used to induce malaise in 0, 1, 2, or 3 mice/cage; all others received saline. Prior to the study, mice were habituated to handling and injections (HI) with positive reinforcement. Nest score (NS), number of mice in the nest, and anhedonia (sugar cereal consumption) were recorded at: baseline, cage change, HI, injection, and injection+cage change. Analyses used GLMs with post-hoc contrasts. Number of mice outside the nest was not affected by any treatment. NS was altered in solitary ($P<0.001$) but not group housed mice ($P=0.15$). LPS injected solitary mice had lower NS than saline mice at injection+cage change ($P<0.001$). LPS mice, in both social treatments, ate ≈2 g less sugar cereal per mouse at both injection and injection+cage change compared to saline controls (2.7-2.8 g; $P<0.05$) and baseline consumption (2.6-3.2 g; $P<0.05$). Group housing appears to mask changes in NS if conspecifics are healthy and acutely ill individuals do not appear to be excluded from the nest. However, a reduction in NS can be used to identify acute illness after cage change, especially in solitary mice. This technique may be useful for earlier identification of illness for general husbandry purposes and may be a more robust tool in chronic disease models.

Effects of depression score on welfare implications of CO_2 and argon gas euthanasia of piglets

Larry Sadler[1], Locke Karriker[1], Anna Johnson[1], Chong Wang[1], Tina Widowski[2] and Suzanne Millman[1]
[1]*Iowa State University, Ames, IA 50011, USA,* [2]*University of Guelph, Guelph, ON, N1G 2W1, Canada; ljsadler@iastate.edu*

Based on 2006 USDA data, preweaning mortality is 1.1 piglets per litter, or 10.9% of all piglets born. Preweaning mortality consists of piglets that die due to disease, crushing or other causes as well as those that are euthanized on farm. CO_2 is an approved euthanasia method for piglets according to the American Veterinary Medical Association, and according to the American Association of Swine Veterinarians, CO_2 is increasingly adopted for piglet euthanasia on commercial swine facilities. Argon is proposed as a more humane alternative, killing through hypoxia vs. hypercapnia. Since the physiologic effects of CO_2 and Argon differ, it is important to examine humaneness of both gases when applied to moribund piglets. The objectives of this study were to compare the effects of depression score, gas type and flow rate on the welfare of piglets during euthanasia. Piglets identified for euthanasia were assessed using a 4 point depression score. Moribund (MOR; depression score 3) and other (OTH; depression score 0 or 1) were enrolled in the study. Gases consisted of 100% CO_2 (n=44 piglets) and 100% Ar (n=16 piglets), administered at prefill and gradual fill (35% chamber volume exchange/min) flow rates. Piglet pairs (1 MOR, 1 OTH) were placed in a Euthanex AgPro™ box, modified with clear plastic on the lid and one side to facilitate behavior observations. Duration of open mouth breathing (OMB) and latencies to loss of posture (LP), last limb movement (LLM) and final gasp response (FG) were collected using direct observation and video. Data were censored at 10 min for LP and 25 min for FG, at which time any piglets displaying these responses were removed from the study and a secondary euthanasia method (blunt force trauma) was applied. Analyses of data were performed in SAS as a linear mixed model. Contrary to our hypothesis, MOR piglets did not differ from OTH when CO_2 was applied at either flow rate (Prefill: mean±SE [seconds(s);P-value] for MOR vs OTH: OMB=14±3 vs. 21±3 [s;0.09]; LP=37±5 vs 40±5 [s;0.74]; LLM=142±50 vs 167±50 [s;0.72]; FG=377±72 vs. 400±72 [s;0.82]). Conversely, in the Ar prefill treatment, MOR piglets displayed longer duration of OMB (126±21 vs 46±21 [s;0.002]), and longer latency to LP (213±32 vs 77±29 [s;0.001]) and LLM (511±72 vs 816±72 [s;0.004]). Ar treatments were discontinued for ethical reasons; all piglets in gradual fill (n=4 piglets) and 50% in the prefill (n=6 piglets) required a secondary euthanasia step. In conclusion, depression score did not affect the welfare of piglets when euthanized using CO_2 at either flow rate. Ar was not an acceptable euthanasia method at either flow rate, particularly for MB suckling piglets.

Does rubber flooring influence postural and spatial behaviour in gestating sows?

Julia Adriana Calderón Díaz[1,2], Alan Fahey[1] and Laura Boyle[2]
[1]*School of Agriculture and Food Sciences, University College Dublin, Belfield, Dublin 4, Ireland,*
[2]*Pig Development Department, Teagasc Animal and Grassland Research and Innovation Centre,*
Moorepark, Fermoy, Co. Cork, Ireland; julia.calderon-diaz@teagasc.ie

This study evaluated the effect of flooring on postural and spatial behaviour of group housed gestating sows. Sixty-four sows were kept in groups of 4 in pens (area available per sow = 2.8 m^2) with solid concrete floored feeding stalls and a concrete fully slatted group area from 28 days post service. The slats were either left uncovered (CON; n=8) or 10mm thick rubber slat mats were affixed (the gaps were uncovered) (RUB; n=8). Sows were video recorded for 24h on 1, 8, 25, 50 and 75 days after entering the trial. Videos were sampled instantaneously every 10min. An index of the proportion of time spent in different postures (Standing [S], Ventral [VL] and Lateral Lying [LL] and total lying [L]), locations (stalls and group area) and posture by location was calculated. Variables were tested for normality and analysed using SAS V9.3 PROC MIXED. There was no association between the time spent in each posture and floor type ($P>0.05$). RUB sows spent more time in the group area (76.3% vs. 53.3±5.8%; $P<0.01$) compared with CON sows. CON sows tended to spend more time in the feeding stalls (45.4% vs. 28.4±5.9%; $P=0.06$) compared with RUB sows. RUB sows stood less (19.1% vs. 35.6±5.0%; $P<0.05$) and lay more (80.1% vs. 62.5±5.2%; $P<0.05$) in the group area than CON sows. Additionally, RUB sows tended to spend more time LL (46.6% vs. 32.8±5.1%; $P=0.08$) in the group area and less time LL (17.6% vs. 38.2±5.6%; $P<0.05$) in the stall area compared to CON sows. Sows spent more time in the group area when it was covered by rubber slat mats. Additionally, RUB sows spent more time lying in the group area compared to sows with no access to rubber flooring. This reflects the preference of group housed sows for a comfortable surface for lying during pregnancy.

How much straw do slaughter pigs need to meet the need for exploration?

Lene Juul Pedersen[1], Mette Herskin[1] and Björn Forkman[2]
[1]Aahus University, Animal Science, Blichers Allé 20, 8830 Tjele, Denmark, [2]Copenhagen University,
Life Science, Grønnegaardsvej 8, 1870 Frederiksberg C, Denmark; lene.juulpedersen@agrsci.dk

According to EU Directive (2001/93/EC) 'pigs must have permanent access to sufficient quantity of material to enable proper investigation and manipulation activities'. By providing rooting materials the need of pigs for exploration are met, and this, in turn, reduces the risk of pigs to redirect explorative behavior towards pen-mates. The question is, however, what constitutes a sufficient quantity? In two experiments, we investigated the relation between redirected explorative behavior towards pen-mates and amount of straw provided daily to pigs (10 to 1000 g/pig/day) to find the amount where additional straw no longer reduced redirected bahviour. Pigs were housed in groups of 18 at 0.7 m^2/pig in pens with 1/3 slatted floor. Pens were cleaned manually twice weekly and fresh long straw was provided daily to each pen in a pile. Duration of all occurrences of redirected explorative behavior towards pen mates (touching another pig with the snout) was observed at 40 and 80 kg during 12 h. Fourty-eight groups were assigned randomly to 10, 500 or 1000 g of straw/pig/day. The results showed a reduction in redirected behavior when pigs were given 500 compared to 10 g ($P=0.04$) but no further reduction when given 1000 g. Hereafter, 96 groups were randomly assigned to 8 treatments (10, 80, 150, 220, 290, 360, 430 or 500 g straw/pig/day). Data were analyzed using mixed models allowing for a curved relationship. The results showed a significant linear relation ($P=0.01$) between straw and redirected behavior with a reduction in time spent with redirected behavior from 10.6 to 8.2% of active time when pigs were provided 500 g compared to 10 g. Thus, the results showed that penmate-directed behavior was gradually reduced with increasing amount of straw, indicating that pigs are not saturated with straw until the amount is close to 500 g of straw/pig/day.

The use of perches by broiler chickens

Paula Baker, Jon Walton, Ibrahim Renan, Joanne Edgar, Justin Mckinstry, Andrew Butterworth and Claire Weeks
University of Bristol, Animal Behaviour and Welfare, Langford, North Somerset, Bristol, BS40 5DU, Bristol, United Kingdom; p.e.baker@bristol.ac.uk

The changes you have again suggested have already been edited throughout the abstract previously. See below after the abstract text changes that have been amended, Thank you. Environmental enrichment has been widely used for companion and zoo animals. Previous studies have shown that environmental enrichment for commercial meat chickens (broilers) can enhance physical activity and improve leg health. Perching is a behavioural need of chickens. This study measured the effect of providing perches for 30,000 mixed sexed Ross birds at 2,3,4 and 5 weeks of age. Birds were reared in two identical houses enriched with two different types of perching objects; 4 wooden A-frame perches consisting of rungs at different heights (15 cm, 30 cm, 45 cm and 60 cm) and 30 chopped straw bales (encased in plastic). Both sets of enrichment were evenly distributed throughout each house. Data were collected by focal observations and consisted of 10 minutes observations over 2 days a week within defined quadrats. Direct observations were used to record the number of birds using the bales and on the different rungs/levels and underneath perches. The number of birds positioned against the bale and on the bale was also taken into account within light and dark areas of the shed. Birds appeared to rest underneath the perches, with the highest proportion 91% (252 birds/278 birds in the quadrats) at 2 weeks of age. Although birds accessed all levels, the lowest rung saw the highest number of birds with 24% (58/242) at 4 weeks. By 5 weeks 3% (7/253) perched on the highest rung and 12% (30/253) on the lowest rung. Data was analysed by Freidman Test and there was overall significant variation between the number of birds on or against the bales in light and dark areas across the different weeks of age ($P=0.003$), but insignificant differences when age of birds and position of bale were considered independently. Broiler utilised all levels of perches at each age. This study demonstrates that providing perching opportunities enabled birds to express normal behaviour.

The impact of space availability in movement patterns of dairy sheep during pregnancy

Xavier Averós[1], Areta Lorea[1,2], Ina Beltrán De Heredia[1], Josune Arranz[1], Roberto Ruiz[1] and Inma Estevez[1,3]

[1]Neiker-Tecnalia, Department of Animal Production, Arkaute Agrifood Campus, P.O. Box 46, 01080 Vitoria-Gasteiz, Spain, [2]Navarra Public University (UPNA), Animal Production Department, Campus de Arrosadia, 31006 Pamplona, Spain, [3]IKERBASQUE, Basque Foundation for Research, Alameda Urquijo 36-5 Plaza, 48011 Bilbao, Spain; xaveros@neiker.net

Freedom of movement is essential to guarantee farm animal welfare. However, animal movement may be compromised by management practices severely restricting space availability. To determine the effect of space availability on the use of space and movement patterns of pregnant dairy ewes (*Ovis aries*), 54 individuals were monitored during the last 11 weeks of pregnancy. Three space availabilities (1, 2, and 3 m^2/ewe) were tested maintaining group size constant at 6 ewes/group. Data were collected 2 days/week, with 2 rounds of observations/day by the same observer. Ewes' positions in XY coordinates were collected during 15 minutes using continuous scan samplings (about 1 scan sampling/ewe/minute) with the Chickitizer software. Parameters calculated from XY coordinates included: total and net travelled distances, net/total distance ratio, maximum and minimum step (distance travelled per ewe between 2 consecutive sampling intervals), movement activity (frequency of XY changes during each 15 minutes period), angular dispersion (a parameter estimating the tortuosity of trajectories), and nearest and furthest neighbour distance. A mixed model ANOVA was used to test the effect of space availability, including week as the repeated measure and enclosure as random factor. Lower total travelled distance, net/total distance ratio, maximum step, nearest and furthest neighbour distance, and angular dispersion were observed at 1 m^2/ewe ($P<0.05$), with the same trend observed for net distance ($P<0.10$). Movement activity was also higher at 1 m^2/ewe ($P<0.05$). Results show a clear reduction in the movement length and steps composing it, a reduction in inter-individual distances, a higher degree of tortuosity of the trajectories and higher restlessness for lower space availability. The major effect found in most of the parameters defining movement and use of space demonstrates that a reduction in space availability below 2 m^2/ewe limits movements. However, both enclosure size and density may have contributed to this effect.

Pekin duck social behavior: time budgets, synchrony and preferential associations

Maja M. Makagon[1,2], *Stephanie S. Robles*[1,2] *and Joy A. Mench*[2]
[1]*Purdue University, Department of Animal Sciences, 125 S. Russell St., West Lafayette, IN 47907, USA,* [2]*University of California, Davis, Department of Animal Science and Center for Animal Welfare, One Shields Ave., Davis, CA 95616, USA; mmakagon@purdue.edu*

Pekin ducks share nest boxes even when a separate box is available for each duck, indicating the potential importance of social influences on resource use in production settings. We evaluated the effects of nest box availability on behavioral synchrony and the formation of preferential associations among ducks, and also whether ducks that nest together associate preferentially in other contexts. Sixty-four female Pekin ducks were housed in groups of 8 with access to either 2 or 8 individual nest boxes (4 pen replicates per treatment). The behavior of the ducks was video recorded on 3 consecutive days at 34 wk of age at 03:00-07:00, 10:00-14:00 and 17:00-21:00h. The numbers and identities of ducks performing the following behaviors were recorded at 15 min intervals: nesting, eating, drinking, resting, comfort, other. Time budgets were calculated for each time period and pen. Synchrony, assessed using the kappa statistic, and preferential association analyses, based on permutation of association matrices, were conducted separately for each behavior, time point and pen. The proportion of time that ducks spent in nest boxes (avg. 30% of morning) did not differ between the 2 treatments (t3 = 0.18, $P=0.87$). However, nesting was only synchronized in pens with 8 boxes, suggesting that nest box occupancy turnover is higher when nest box availability is limited. Although nesting synchrony within the 8 box groups could occur without nest sharing, the preferential association analyses confirmed that sharing was occurring. Up to 6 (range 1-6) preferentially associating dyads were identified in pens with 8 boxes, but there were typically few such associations in pens with 2 boxes (range 0-5). Dyads that preferentially shared nest boxes did not preferentially associate in other contexts (ex. resting). These results demonstrate that social influences affect nesting behavior, but that this effect is mitigated by resource availability.

The temperament of peccaries (*Mammalia, Tayassuidae*) applied to their management in captivity

Selene Nogueira[1], Jaqueline Macedo[1], Aline Santánna[2], Sérgio Nogueira-Filho[1] and Mateus Costa[2]
[1]Universidade Estadual de Santa Cruz, Laboratório de Etologia Aplicada, Rod. Jorge Amado s/n km 16, Salobrinho, Ilhéus, 45662900, Brazil, [2]Universidade Estadual Paulista- Jaboticabal, Pós-Graduação em Genética e Melhoramento Animal-FCAV, Prof. Paulo Donato Castellane s/n, 14884-900, Brazil; seleneuesc@gmail.com

The captive breeding of white-lipped peccary (*Tayassu pecari*) – WLP – and the collared peccary (*Pecari tajacu*) – CP – is indicated as an alternative to overhunting and deforestation in Neotropical countries. Both species, however, show some wariness during handling, which may result in poor animal welfare. Thus, we compared the response and temperament of 17 WLPs and 18 CPs when restrained for 20s in a chute, to evaluate the animals' reactions to management procedures. We compared the time spent to drive the animals into the chute and their exit speed after being released from the chute. Their temperament was assessed when the peccaries were restrained in the chute using the qualitative behavior assessment method (QBA), considering 15 adjectives as descriptors of peccaries' temperament. The mean scores of four judges (70% of reliability accordance) on the QBA data were analyzed through principal component analysis (PCA). We used Mann-Whitney U-test in all comparisons between WLP and CP. There was no difference in the time spent to drive both species into the chute (CP: 26.1±7.3s; WLP: 44.9±19.1 s; U=121.0; df=17; P=0.2). However, WLPs were faster than CPs in exiting the chute (WLP: 2.3±0.5m/s; CP:1.5±0.5 m/s; U=85.5; df=17; P=0.04). The first principal component explained 80.7% of the temperament data variance, with the highest negative loadings for the adjectives relaxed, calm, bored, apathetic, satisfied and docile, and the highest positive loadings for the adjectives aggressive, active, fearful, agitated, tense, alert, nervous, anxious and stressed. The CPs showed lower scores than WLPs in the first principal component (CP: -0.3±0.2; WLP: 0.4±0.5; U=90; df=17; P=0.02). Therefore, apparently, CPs showed more behavioral predisposition to be tamed and farmed. WLP reactivity revealed that we need to improve WLP management practices, comparing, for example, the use of different shapes of corral-traps to decrease animals' stress, improving their welfare. This work was suported by PROCAD-NF/CAPES 794-2010. Sérgio Nogueira-Filho and Selene Nogueira were supported by CNPq (Process # 300587/2009-0 and 306154/2010-2, respectively). JFM was supported by CAPES.

Behavioral response of guanacos (*Lama guanicoe*) to tourists in Torres del Paine National Park, Chile

Beatriz Zapata[1], Nicolas Fuentes[2], Benito Gonzalez[2], Juan Traba[3], Cristian Estades[2], Pablo Acebes[3] and Juan Malo[3]
[1]Universidad Mayor, Escuela de Medicina Veterinaria, Camino La Piramide 5750, Chile, [2]Universidad de Chile, Laboratorio de Ecologia de Vida Silvestre, Facultad de Ingenieria Forestal, Santa Rosa 11315, Chile, [3]Universidad Autonoma de Madrid, Terrestrial Ecology Group-TEG. Departamento de Ecología. Facultad de Ciencias, C/ Darwin, 2. E-28049 Madrid, Spain; beatrizzapata@hotmail.com

Tourism in protected wildlife areas may have detrimental effects on inhabitant animals; consequently it is important to continuously monitor their response to visitors. Torres del Paine (TdP) National Park has an important population of guanaco which has been exposed to a growing touristic activity since this reserve was created. Because humans can be perceived as predators by animals, we hypothesize that guanacos have behavioral changes when human approach them, decreasing foraging and increasing vigilance time. This behavioral response could be modulated by habituation to human presence. In order to test this hypothesis, frequency and time assigned to vigilance, foraging and restlessness (number of behavioral changes) were measured in three occasions: (1) a hidden position, 200 m away from the centre of the group; (2) one experimental tourist approached to 100 m; and then (3) to 50 m. The procedure was conducted inside (high tourism) and outside (low tourism) TdP and vehicle traffic (VT) was also considered. We observed family groups, continuously recording behavior of the dominant male and three females randomly chosen, and performed scan sampling of the whole group to record foraging and vigilance rate. Repeated measure ANOVA was conducted, using distance of experimental tourist and location (inside/outside) as independent variables. VT index and group size were covariates. Dependent variables were restlessness and percentage of time foraging and vigilance. Results indicated that distance between guanacos and experimental tourist did not significantly affect their behavior; vigilance was only affected by group size ($F(1,32)=4.23$, $P=0.048$). Location and sex had a significant interaction, where vigilance was greater in males outside TdP ($F(1,31)=7.3$; $P=0.011$) and restlessness increased as VT index was higher (females: $F(1,32)=4.3$; $P=0.046$; males: $F(1,31)=5.9$; $P=0.021$). We concluded that guanacos in TdP are highly tolerant to humans; behavioral changes observed could be explained by current human activities (location/VT) rather than by the presence of tourists.

A multi-zoo study on the effect of manipulated visitor behaviour on zoo meerkats

Sally Sherwen[1], Michael Magrath[2], Kym Butler[1,3], Keven Kerswell[1] and Paul Hemsworth[1]
[1]*The University of Melbourne, Animal Welfare Science Centre, Victoria 3010, Australia, [2]Zoos Victoria, Wildlife Conservation and Science, Melbourne Zoo, Parkville 3052, Australia, [3]Future Farming Systems Research, Biometrics Unit, Department of Primary Industries, Victoria 3030, Australia; sherwens@unimelb.edu.au*

The literature indicates that visitors may affect both the behaviour and welfare of zoo animals. Meerkats, *Suricata suricatta*, at three exhibits were studied under two treatments: (1) unregulated visitor behaviour; and (2) regulated visitor behaviour. The regulated treatment consisted of the presence of a zoo official and signage requesting visitors to be quiet and not to interact with animals. At each exhibit, treatments were imposed using a 4-replicate paired comparison design, with each pair consisting of two consecutive days of different treatments. Meerkat behaviour (e.g. vigilance, resting) and location were recorded using instantaneous sampling every 2 minutes over a total of 72 h across exhibits. The efficacy of the regulated treatment in moderating visitor behaviour was evaluated by recording visitor noise using a decibel logger and by ranking intensity of visitor behaviour on a 0-2 scale (from passively observing to actively attempting to gain the animals' attention) every 2 minutes throughout the study days. Treatment effects were evaluated using analyses of variance with separate strata for exhibits, day-pairs and days. The regulated treatment was successful in reducing visitor noise by 4dB (sed=0.6) at each exhibit ($P=0.0001$) and intensity of visitor behaviour (from 0.69 to 0.12, 0.20 to 0.05 and 0.48 to 0.04 for the three exhibits, main effect $P=0.00001$). However, despite good experimental precision, the regulated treatment did not change the distance meerkats positioned themselves from visitors (3.1 and 2.9m, sed=0.15, $P=0.22$), or the proportion of time engaged in vigilant behaviour (transformed means (back transformed) 34 (0.32) and 35 (0.34), sed=1.9, $P=0.56$) and resting behaviour (transformed means (back transformed) 20 (0.11) and 13(0.05), sed=4.8, $P=0.18$). This study provides experimental evidence that meerkats are behaviourally unreactive to the intensity of visitor behaviour, at least within the range of typical zoo visitor interactions at these sites.

Evaluation of rubber rings coated with lignocaine for pain relief during castration of lambs

Mairi Stewart[1], Craig B Johnson[2], James R Webster[1], Kevin J Stafford[2] and Ngaio J Beausoleil[2]
[1]AgResearch Ltd, Innovative Farm Systems, Private Bag 3123, Hamilton 3240, New Zealand, [2]Massey University, Institute of Veterinary, Animal and Biomedical Sciences, Private Bag 11-222, Palmerston North 4442, New Zealand; mairi.stewart@agresearch.co.nz

To facilitate wider use of pain relief on-farm, 'farmer friendly' methods for pain relief are necessary. We evaluated the efficacy of rubber rings coated with local anaesthetic (LA) for providing pain relief in 54 lambs (4 weeks old) randomly allocated to either: (1) handling without castration (C); (2) castration with a normal rubber ring (R); (3) castration with a rubber ring coated with LA (lignocaine, RLA). To eliminate any potential effects of blood sampling on behavioural responses, the study was carried out in two parts utilizing different animals. Behavioural responses were measured for 30 lambs (n=10 per treatment) and on the following day, cortisol responses were measured for 24 lambs (n=8 per treatment). Frequency of behaviour (abnormal/normal lying, restlessness, transitions standing to lying) was recorded by video for a 30-min baseline period and 3 hours post-treatment. Blood samples for cortisol were taken via jugular venipuncture at 0, 30, 60, 90 and 120 min. REML was used to detect treatment differences. Transitions from standing to lying, restlessness and abnormal and normal lying increased from baseline for R and RLA ($P<0.001$), however, transitions from standing to lying and abnormal lying were less frequent ($P<0.05$) for RLA within one hour post-treatment. The total cortisol response was greater for R than RLA ($P<0.05$). Cortisol peaked 60 min after treatment for both R and RLA and was lower for RLA than R ($P<0.05$) at 90 and 120 min. The lower cortisol and behavioural responses in the RLA lambs indicated that the lignocaine coated ring reduced some of the pain associated with castration; however diffusion through the skin appeared delayed. Therefore, this technique has potential as an easy, practical way of administering lignocaine and provides a 'farmer friendly' method for administering pain relief on-farm. Further development is required to achieve faster absorption into the tissue and improve pain alleviation.

Achieving a high quality of life for farm animals requires provision, rather than deprivation

Rebecca Sommerville and Tracey Jones
Compassion in World Farming, Food Business, River Court, Mill Lane, Godalming, Surrey, GU7 1EZ, United Kingdom; Rebecca.Sommerville@ciwf.org.uk

The Five Freedoms are internationally recognised to define acceptable states of animal welfare. However, there are many instances when they are not met in practice, as farming systems control and restrain animals to manage health, high productivity, and harmful behaviours. Control is largely achieved through some form of deprivation, which denies one or more of the Freedoms. Deprivations include lack of space, natural light, environmental complexity, social contact, feed, or a physical appendage. Animals are sentient beings capable of feeling emotions such as 'fear' or 'pleasure'. Depriving them of basic needs deprives them of an acceptable welfare state. We suggest 'provision rather than deprivation' is key to achieving a high quality of life for farm animals. We present four examples where deprivations are common, but research indicates provision is better for welfare. (1) To control feather pecking in laying hens, their beaks are partially amputated (trimmed). Research shows that providing hens with a complex environment with opportunity to forage, peck and scratch reduces the risk of feather pecking, allowing beaks to remain intact. (2) To control piglet mortality, sows are kept in farrowing crates indoors, depriving them movement and the ability to perform nesting and maternal behaviour. Systems are being developed that allow sows behavioural expression, without escalating piglet mortality. (3) Dairy calves are individually housed, largely to control enteric disease, depriving them social contact. Research shows that group housing can improve growth and social confidence. (4) Inactivity in broiler chickens results from high growth rates and barren environments. Reducing growth rate and providing broilers with natural light and a complex environment can improve health and activity levels. How provisions fulfil the physical, psychological and behavioural needs of animals whilst remaining viable in commercial practice is considered.

Piagetian object permanence abilities in goat kids

Rachel Chojnacki, Judit Vas and Inger Lise Andersen
Norwegian University of Life Sciences, Department of Animal and Aquacultural Sciences, P.O. Box 5003, 1432 Aas, Norway; rachel.chojnacki@umb.no

Object permanence is the understanding that an object continues to exist when removed from the plane of vision. According to Piaget, the understanding of object permanence develops in six stages (1-6, with stages 4, 5 and 6 having two sublevels). Piaget`s object permanence concept has been used to assess the cognitive abilities of a variety of animal species as the ability to hold a mental image of an object is considered to require advanced cognitive capabilities. Object permanence tasks, to our knowledge, have previously not been investigated in farm animal species. Previous behavioral studies have demonstrated that goats are able to perform learning and memory tasks which are considered to require higher brain functions; therefore, we studied the object permanence abilities in goat kids. Using a milk bucket with artificial teats, a blind and two hiding curtains, 36 Norwegian dairy goat kids, whose mothers had been held at treatment densities of 1, 2 and 3 m^2 throughout pregnancy, were tested. The object permanence tasks assessed their cognitive aptitudes at four levels (three visible and one invisible displacement problems) from Piagetian stages 4a to 6a at six weeks of age. Ten trials were administered at each level over the course of 4 days. The criterion for passing each stage was set at 80%. According to the binomial test, the goat kids were able to reach the success criterion at each stage and advance onward to perform stage 6a of Piagetian object permanence tasks significantly above chance ($P=q=0.5$, a=.05, n=36, 22, 14 and 14 at stages 4a, 4b, 5a and 6a, respectively). No effect of the prenatal treatment on the kids' abilities to perform the object permanence tasks was found (p-value=0.72). The results suggest that goats are capable of higher cognitive functioning and are able to understand and perform advanced cognitive tasks.

Temperamental turkeys: do tonic immobility, open field and novel object test responses change?

Marisa Erasmus and Janice Swanson
Michigan State University, Animal Science, 474 S. Shaw Lane, East Lansing, Michigan 48824, USA; erasmusm@msu.edu

Tonic immobility (TI), open field (OF), and novel object (NO) tests have traditionally been used to assess fear responses of poultry, and can also assess activity levels and coping styles. These tests are often used to draw conclusions about well-being (e.g. the NO test is part of the Welfare Quality® Assessment Protocol for Poultry). Few studies have examined how turkeys respond to these tests. More importantly, it is unknown whether turkeys' responses change over time. Male turkeys were housed in groups of 4-6 in 16 pens for 14 wk. Turkeys (n=80) were tested in all three tests (balanced for order) between 4 and 6 wk (period 1), and between 8 and 10 wk (period 2). Measures recorded during testing included latency to vocalize (TI and OF tests); whether or not vocalizations occurred (Y/N, occurrence), and number of vocalizations (TI and OF tests); TI duration; and latency to ambulate and number of areas entered (OF test). NO test measures included whether or not birds approached and pecked the NO (Y/N), latency to move within one body length, latency to approach, and latency to peck the NO. Results were analyzed using the GLIMMIX and LIFETEST procedures in SAS. Duration of TI and latency to vocalize did not differ between TI test periods. The number of vocalizations did not differ between periods for TI or OF tests. The occurrence (Y/N) of vocalizations ($P=0.04$), latency to ambulate ($P=0.02$), and latency to vocalize ($P=0.0008$) differed between OF test periods, but the number of areas entered did not. All NO test measures differed between periods ($P<0.03$). This study is the first to show that some fear-related behavioral responses of turkeys change over time, especially during NO testing. Caution is needed when interpreting some of these test measures in relation to well-being.

Temperament traits of sows in two group housing systems

Yolande M Seddon[1], Jennifer A Brown[1], Fiona C Rioja-Lang[1], Nicolas Devillers[2], Laurie Connor[3] and Harold W Gonyou[1]

[1]Prairie Swine Centre, Department of Ethology and Welfare, 21057, 2105 8th Street East, S7H 5N9, Canada, [2]Agriculture and Agri-Food Canada, Dairy and Swine Research and Development Centre, 2000 College Street, Sherbrooke, Quebec, J1M 0C8, Canada, [3]University of Manitoba, Department of Animal Science, Winnipeg, Manitoba, R3T 2N2, Canada; yolande.seddon@usask.ca

The interaction between temperament and housing may have important consequences for the suitability and productive success of sows in group housing. This study evaluated temperament of sows housed in two group systems: electronic sow feeders (ESF) with partial slatted floors (CONV, n=138 sows), or ESF with deep straw bedding (ALT, n=146 sows), to determine if differences in temperament existed and whether this was linked to productivity. Sows were of Genesus genetics, with parity ranging from 0-7. Systems were identical in layout and group size varied between 21-30 sows. Temperament was assessed in all sows at eight weeks of gestation using four behavioural tests: (1) open door test (ODT); (2) novel object test (NOT); (3) human approaching pig test (HAP); and (4) pig approaching human test (PAH), with the goal of assessing active/passive and confident/fearful traits. Effects of housing system, age and breeding cohort on behaviour and productivity measures were determined using PROC MIXED (SAS, 9.2). Data were then reanalysed separately by housing system and residuals from behaviour measures were subjected to factor analysis (PROC FAC) to determine temperament characteristics for each sow. Temperament factor loadings were correlated to the residuals of productivity measures. ALT and CONV sows differed in their response to ODT, NOT and HAP tests ($P<0.005$); with ALT sows showing more active and fearful traits. A higher body injury score (BIS) measured at 16 weeks gestation was positively correlated to active traits in CONV ($P<0.005$), but not in ALT sows. Rather, in ALT sows, the BIS was positively correlated to fearful traits ($P<0.05$). In ALT, but not CONV, active traits correlated to a greater backfat loss over gestation ($P<0.005$). Housing system clearly influences the behavioural responses and temperament traits of group-housed sows, and should be considered when developing selection and management strategies for sows.

Identification of temperamental categories in horses

Malin Axel-Nilsson[1], Kathalijne Visser[2], Sara Nyman[3], Kees Van Reenen[2] and Harry Blokhuis[1]
[1]*Swedish University of Agricultural Sciences, Department of Animal Environment and Health, Box 7068, 750 07 Uppsala, Sweden,* [2]*Wageningen UR, Livestock Research, Lelystad, the Netherlands,* [3]*Wallby Säteri, Skirö, 574 96 Vetlanda, Sweden; Malin.Axel-Nilsson@slu.se*

It is generally recognized that horse temperament is among the main factors affecting the performance of a horse and rider combination. However, horses sometimes show a display of temperamental characteristics that may be difficult and dangerous to handle for the non-professional rider or handler. It may also lead to impaired welfare of the horse if its temperament doesn't match the responsible person's own personality, experience and knowledge. There is no well-established way to test and evaluate temperament in horses. The aim of this study was to investigate the use of existing behavioural tests to identify temperamental categories in horses. The study included seventeen Swedish Warmblood horses and combined four tests; Human Approach, Isolation, Handling and Novel Object, that were performed twice. Both heart rate measurements and behavioural observations were made. Response measures that exhibited consistency over repeated tests (Spearman Rank Correlation coefficients, $P<0.05$) were averaged across repetitions, and the averages were subjected to principal component analysis (PCA) in two steps. First, four separate PCA analyses on measures recorded from the same test were conducted, followed by PCA on scores of the first two principal components of each first-step PCA (8 components in total). Two summary factors (eigenvalue >1), explaining 49% of the variance, were obtained and used to categorize the sample of horses in groups; high responders and low responders. In terms of loadings on the original measures, factor 1 had high loadings for walking (+), exploration (+), standing (-) and snorting (-) during the Novel Object test, whereas factor 2 had high loadings for heart rate (+), cantering (+), and bucking/kicking (+) during Novel Object as well as high loading for walking (-) during the Isolation test. The results indicate that responses to behavioural tests are influenced by multiple underlying temperamental traits, and can be used to categorize horses.

The evaluation of behavioural and clinical parameters in assessing weaner pig recovery on farm

Rebecca L Wilson[1], Patricia K Holyoake[2], Greg M Cronin[3] and Rebecca E Doyle[1]
[1]Charles Sturt University and EH Graham Centre, Locked Bag 588, Wagga Wagga, NSW, 2678, Australia, [2]Department of Primary Industries, Cnr Midland Hwy and Taylor Street, Bendigo, VIC, 3100, Australia, [3]University of Sydney, 425 Werombi Road, Camden, NSW, 2570, Australia; rewilson@csu.edu.au

Improved animal welfare in commercial settings relies on early identification and treatment of ill animals. This experiment looked at common on-farm measures of welfare as well as infrared eye temperature (IET), to assess recovery rates of weaner pigs in hospital pens. The experiment was conducted on farm using 136 weaner pigs (3-5weeks). Ill or injured pigs were matched for gender and illness upon admittance to a hospital pen, and the assessor was unaware of the treatment groups. All pigs were administered an antibiotic suitable for their condition and either meloxicam (0.02 mL/kg) or ketoprofen (0.03 ml/kg) daily for three days. Health was assessed daily by rectal temperature (RT), IET (image taken 45 cm from eye, temperature measured using dot point analysis), activity score (number of times found lying compared to sitting, standing or kneeling; every 15 min from 12:00 to 07:00 h on days 1 and 4), and body weight change (BW; taken on days 1 and 4). Behavioural data were analysed using binomial GLMM, and all other measures with LMM. There was a significant difference between treatments for activity, with the ketoprofen treatment group spending less time overall (87±0.12%) lying than pigs treated with meloxicam (90±0.12%; $P=0.006$). A difference in RT between day 1 (39.78±0.06 °C) and all other days, (39.53±0.06 °C, 39.45 ±0.06 °C, 39.51±0.06 °C; $P<0.001$) was evident, as was a difference in IET between days 1 (33.91±0.21 °C) and 3 (34.34±0.21 °C; $P<0.001$). BW did not change over the trial ($P=0.971$). Results showed that activity score may be the most useful method to detect recovery rates and indicated that infrared eye temperature may be sensitive enough to reveal small changes in temperature indicating illness, but further investigation is required.

Differences between female and castrated male pigs in their response to novelty

Inonge Reimert[1], T. Bas Rodenburg[2], Winanda W. Ursinus[1,3], Bas Kemp[1] and J. Elizabeth Bolhuis[1]
[1]Wageningen University, Adaptation Physiology Group, De Elst 1, 6708 WD Wageningen, the Netherlands, [2]Wageningen University, Behavioural Ecology Group, De Elst 1, 6708 WD Wageningen, the Netherlands, [3]Wageningen UR Livestock Research, Animal Sciences, Animal behaviour & Welfare, Edelhertweg 15, 8219 PH Lelystad, the Netherlands; inonge.reimert@wur.nl

As behavioural differences between male and female pigs could have implications for their welfare, differences between female and castrated (at 3 days of age with CO_2/O_2 anesthesia) male pigs in their response to novelty were investigated. Before weaning (bf), 80 litters were litter-wise subjected to a 10 min novel object (feeder with feed) test (NOT) and to a 10 min human approach test (HAT) in the farrowing pen and individually to a 2.5-min novel environment test (NET) outside the farrowing pen room. After weaning (af), 240 barrows and 240 gilts were selected and housed with three barrows and three gilts per pen and subjected to a group-wise 10-min HAT in the home pen and individually to a 10 min NET where after 5 min a bucket was lowered from the ceiling. Data were analysed using mixed linear models. Before and after weaning many gender differences were found: e.g. NOT (touch feeder: female: 162.8±9.3 vs. male: 209.2±10.0 s, $P<0.001$), HATbf (touch person: 198.0±9.7 vs. 229.0±10.0 s, $P<0.001$), NETbf (grunting: 52.8±1.9 vs. 57.0±1.9 times, $P=0.04$, squealing and screaming: 3.0±0.5 vs. 4.7±0.8 times, $P=0.02$, defecating: 0.6±0.06 vs. 0.7±0.06 times, $P=0.08$, urinating: 4.8±1.3 vs. 11.8±2.0% of pigs, $P<0.01$), HATaf (approach person: 6.7±0.8 vs. 8.1±0.9 s, $P=0.02$, touch person: 15.9±1.3 vs. 18.2±1.4 s, $P=0.07$), NETaf (touch bucket: 385.2±6.4 vs. 417.1±7.6 s, $P<0.01$). These results suggest that female pigs are less fearful than castrated male pigs which could be due to a difference in early life experience or could be a true gender difference.

Pigs make up their mind early in life: behaviour in novelty tests and blood serotonin

Winanda W. Ursinus[1,2], Cornelis G. Van Reenen[1], Inonge Reimert[2], Bas Kemp[2] and J. Elizabeth Bolhuis[2]
[1]Wageningen UR Livestock Research, Department of Animal Sciences, Animal behaviour & Welfare, Edelhertweg 15, 8219 PH Lelystad, the Netherlands, [2]Wageningen University, Department of Animal Sciences, Adaptation Physiology Group, De Elst 1, 6708 WD Wageningen, the Netherlands; nanda.ursinus@wur.nl

Behavioural responses toward novelty are diverse among individual pigs and type of events. Some behavioural characteristics may be stable throughout life and relate to serotonergic parameters. Our aim was to assess the stability of behavioural responses in a novel environment and to evaluate its relation with blood serotonin. Pigs (n=451, in 5 rounds) were individually exposed to a novel environment at 3 (2.5 min) and 13 weeks of age (5 min; first 2.5 min considered). Behaviours observed were time spent on walking, standing, standing alert, exploring, and no. of vocalizations. Weaning occurred at 4 weeks. Thereafter, pigs were housed barren 'plus' (B+; chain with ball, jute bag, minimal wood shavings) or enriched (E; chain with ball, jute bag, ample wood shavings, straw bedding). Whole blood and platelet serotonin levels were determined in week 8 and 22. Pearson correlations were performed on residuals of a GLM with round as effect. Vocalizations expressed in week 3 were correlated with vocalizations at 13 weeks (B+: r=0.29, $P<0.0001$; E: r=0.33, $P<0.0001$). In B+ pigs, exploration was correlated (r=0.22, $P<0.001$), while in E pigs walking (r=0.26, $P<0.0001$) was correlated between the tests. Exploration in the first test was, in B+ pigs, correlated with walking (r=0.23, $P<0.001$) and standing alert (r=-0.21, $P<0.01$) in the second test. In E pigs, whole blood serotonin of week 8 was correlated with vocalizations (r=0.22, $P<0.001$), whereas platelet serotonin in B+ pigs measured in week 8 was correlated with standing (r=0.20, $P<0.01$) and standing alert (r=-0.21, $P<0.01$) in the first test. Individual behavioural responses to a challenge, especially vocalizations, seem moderately stable throughout life. Serotonin was related to behaviours observed in the first test. To conclude, behavioural responses may be stable, but are not consistently related to blood serotonin. Further research investigates the predictive value of behavioural responses and serotonin to perform maladaptive behaviours.

Effects of alginate and resistant starch on feeding patterns, behavior and growth in growing pigs

Carol Souza Da Silva[1,2], Guido Bosch[1], J. Elizabeth Bolhuis[2], Lian J. N. Stappers[2], Hubert M. J. Van Hees[3], Walter J. J. Gerrits[1] and Bas Kemp[2]
[1]Wageningen University, Animal Nutrition Group, P.O. Box 338, 6700 AH Wageningen, the Netherlands, [2]Wageningen University, Adaptation Physiology Group, P.O. Box 338, 6700 AH Wageningen, the Netherlands, [3]Nutreco Research and Development, P.O. Box 220, 5830 AE Boxmeer, the Netherlands; carol.souza@live.nl

Fibers' satiating properties may reduce hunger and improve welfare of restrictedly-fed pigs, however, a reduced energy intake of fiber diets may reduce pigs' growth performances. We assessed the long-term effects of a gelling fiber promoting satiation (alginate, ALG) and a fermentable fiber promoting satiety (resistant starch, RS) on feeding patterns, behavior and growth of growing pigs fed *ad libitum* during 12 weeks. A control diet containing 40% digestible starch was formulated, from which ALG, RS and RS+ALG diets were formulated by exchanging fibers for digestible starch in a 2×2 factorial arrangement. A total of 240 pigs in 40 pens (6 pigs/pen) were used in 2 batches. From all visits to an electronic feeder, the daily and cumulative feed intake (FI), FI per meal, meal duration, inter-meal interval, and number of meals per day were calculated. Pigs were observed in their home-pen for 6 h using 4-min instantaneous scan sampling in week 12. Data were analyzed with a mixed model including ALG (yes/no), RS (yes/no), their interaction, gender and batch as fixed effects, and pen nested within (ALG RS batch) as random effect. ALG increased daily and cumulative FI ($P<0.05$). RS increased FI per meal, meal duration ($P<0.05$) and inter-meal interval ($P=0.05$), and decreased number of meals per day ($P<0.01$). RS+ALG decreased standing and walking, aggressive, feeder-directed and drinking behaviors (RS×ALG interaction, $P<0.05$) as compared with ALG. Growth did not differ between diets. In conclusion, ALG-fed pigs compensated for the reduced dietary energy content by increasing their FI, and achieved similar growth as control-fed pigs, despite increased physical activity. RS-fed pigs changed feeding patterns, but did not increase FI, and yet achieved similar growth as control-fed pigs. These results show that the lower energy content of fiber diets and their influence on satiety do not necessarily reduce pigs' growth performances.

The effect of level of social contact in dairy calves on behavior and health

Margit Bak Jensen[1] and Lars Erik Larsen[2]
[1]*Aarhus University, Department of Animal Science, Blichers Allé 20, 8830 Tjele, Denmark,*
[2]*Technical University of Denmark, National Veterinary Institute, Bülowsvej 27, 1870 Frederiksberg C, Denmark; MargitBak.Jensen@agrsci.dk*

Social contact has previously been shown to improve social skills, but health concerns have also been raised. In previous studies the level of social contact in individual housing has varied being auditory, visual, or physical contact. It is unclear how these compare to each other and to pair housing as regards effects on behaviour and health. To investigate the effect of increasing level of social contact on social behaviour and health, one-hundred-and-ten Holstein calves (50 males, 60 females) in 11 blocks were paired according to birth date and within 60 h of birth allocated to one of five treatments (all 2.25 m^2/calf); individual housing with auditory contact (I), individual housing with auditory and visual contact (V), individual housing with auditory, visual and physical contact (E), auditory and visual contact the first 2 weeks followed by pair housing (VP), or pair housing (P) until the age of 8 weeks. Individual housing with auditory contact (I) corresponds to commercial situations where calves are housed in closed pens or hutches with no calves housed opposite and thus no visual contact to other calves. This treatment provides a control for minimal social contact to address the question if mere visual contact has any effects. At 6 weeks of age calves were subjected to a social test. The higher the level of social contact the shorter the latency to sniff the stimulus calf (square root transformed means ±SE (back transformed mean (s) in brackets): 22.6a±1.50 (511), 18.2b±1.52 (331), 17.9b±1.48 (320), 14.7c±1.48 (216), 14.5c±1.53 (211)for I, V, E, VP and P, respectively, *P*<0.001) and the longer sniffing head during the 10 min test (0.46a±0.31 (0.21), 0.97ab±0.32 (0.94), 1.35b±0.31(1.82), 1.49b±0.31(2.22), 1.77b±0.32(3.13), *P*=0.03). When the youngest calf of a block was 50 days old, one calf from each treatment were introduced to a group pen (2.7 m^2/calf) and calves' behaviour was recorded during 6 h. Calves with a higher level of social contact spent less time butting other calves (9.40a±1.16 (88.4), 11.66a±1.18 (136.0), 10.56a±1.14 (111.5), 4.55b±1.15 (20.7), 2.97b±1.18 (8.8), *P*<0.001). Health checks, conducted every second week, showed in week 6 fewer pair housed calves with normal faeces score (13/22, 11/22, 15/22, 5/22 and 7/22 for treatments I, V, E, P and VP, *P*=0.01). No effect of treatment was found for respiratory scores in any week. The results suggest that calves housed individually were more fearful than pair-housed calves, and that only pair-housing stimulated the development of social skills. Minimal effects on health of allowing physical contact between calves were detected.

Rearing substrate affects play behaviour in an arena test in dairy calves

Mhairi A. Sutherland, Gemma M. Worth, Mairi Stewart and Karin E. Schütz
AgResearch Ltd, Ruakura Research Centre, Hamilton 3240, New Zealand;
mhairi.sutherland@agresearch.co.nz

The pre-weaning rearing environment of dairy calves, such as flooring substrate, can influence health and welfare of calves. The objective of this study was to investigate the effect of rearing calves on two different substrates at three different space allowances on play behaviour in an arena test outside the rearing pen. At 1 week of age, 72 calves were moved into experimental pens (n=4 calves/pen). Pens had floors covered with either river stones (RS) or sawdust (SD). For each substrate type calves were reared at one of three space allowances: 1.0, 1.5 or 2.0 m^2/calf (3 pens/space allowance/substrate). Calves were tested individually in an arena test at 3 and 6 weeks of age. The arena (3.0×15 m) had bark chip flooring, a substrate that was novel to all calves, and the walls were covered with black plastic to prevent calves from seeing the observers or other calves. Calves were moved from their home pen to the arena where behaviour was recorded continuously for 20min. Running duration and the frequency of jumps, kicks and buck kicks were scored. Data were analysed by ANOVA. Overall, during the arena test, calves reared on RS performed more jumps (9.4 vs 6.8±1.13 (means±SEM), RS and SD respectively, $P<0.05$), kicks (3.8 vs 2.3±0.47, RS and SD respectively, $P<0.005$) and buck kicks (13.1 vs 7.1±1.40, RS and SD respectively, $P<0.001$) and spent more time running (2.5 vs 1.9±0.14 min, RS and SD respectively, $P<0.001$) than calves reared on SD. Space allowance had no effect ($P>0.05$) on performance of running or play behaviour. The increased motivation to play by calves reared on RS outside their rearing environment may be related to the level of play these calves performed in their home pens, which was influenced by characteristics of the RS.

Time spent eating solid feed predicts intake in milk fed dairy calves

Joao H. Cardoso Costa, Rolnei R. Dáros, Marina A.G. Von Keyserlingk and Daniel M. Weary
University of British Columbia, Animal Welfare Program, 2357 Main Mall, Vancouver, BC, V6T 1Z4, Canada; jhcardosocosta@gmail.com

The objective of this study was to compare data obtained using continuous video observation of feeding behaviors with intake of solid feed in milk fed dairy calves, and to determine if the relationship varies with calf age. Single-housed dairy calves (n=10) were observed twice, at 2 weeks (14±1 days) and at 7 weeks (48±3 days) of age. The calves were provided 8L of milk/day for 4 weeks and then reduced to 6 l of milk/day from 5 to 8 weeks. Calves were also provided *ad libitum* access to two solid feeds: calf starter and total mixed ration (TMR). Intakes were measured daily. Continuous time-lapsed video was used to determine time spent eating TMR and starter, where eating was defined by the calf's head within the feed bucket. Individual feeding bouts were summed to determine total daily feeding time. Between week 2 and 7, calves increased feed intake (0.02 vs. 1.01 kg/day of starter and 0.02 vs. 2.72 kg/day of TMR), the number of feeding bouts (15 vs. 39 bouts/day for starter and 11 vs. 99 bout/day for TMR) and feeding bout duration (20.9 vs. 32.9 s/bout for starter and 15.3 vs. 115.5 s/bout for TMR). Feeding behavior data obtained from video were closely correlated with daily intakes of starter (R^2=0.92) and TMR (R^2=0.87) at 7 weeks; these relationship were less strong at 2-weeks of age (starter R^2=0.51; TMR R^2=0.13). These results show that solid feed intake, meal frequency and meal duration increase between weeks 2 and 7, and that video recordings of feeding behaviour can provide reasonable estimates of TMR and starter intake at 7 weeks of age.

The effect of competition for teat access on feeding patterns of dairy calves

Emily Miller-Cushon[1], Renée Bergeron[2], Ken Leslie[3], Georgia Mason[4] and Trevor Devries[1]
[1]University of Guelph, Kemptville Campus, Animal and Poultry Science, 830 Prescott St.,
Kemptville, ON, K0G 1J0, Canada, [2]University of Guelph, Campus d'Alfred, Animal and Poultry
Science, 31 St Paul St., Alfred, ON, K0B 1A0, Canada, [3]University of Guelph, Population Medicine,
50 Stone Rd. E, Guelph, ON, N1G 2W1, Canada, [4]University of Guelph, Animal and Poultry
Science, 50 Stone Rd. E, Guelph, ON, N1G 2W1, Canada; emillerc@uoguelph.ca

This study examined how reduced access to teats impacts the development of milk meal patterns and social behavior of dairy calves. Twenty Holstein bull calves were randomly paired, housed together, and provided milk replacer (MR) *ad libitum* via either: (1) 2 teats/pen (non-competitive feeding; NCF); or (2) 1 teat/pen (competitive feeding; CF). Calves were followed for 6 weeks. Feeding times and competitive interactions were recorded from video for 3 d in each of weeks 2, 4, and 6. Meal criteria were calculated to determine daily meal frequency, meal time, and synchronized meal time (defined for both treatments as time when calves within a pen were engaged in simultaneous meals, with meals defined by individual meal criteria). Data were summarized by pen and analyzed in a repeated measures general linear mixed model. Milk intake was subject to a treatment by week interaction ($P=0.002$), with calves in CF pens increasing intake over time to a greater extent than calves in NCF pens (wk 2, 8.3 vs. 9.6 L/calf/d, SE=0.6, and in wk 4-6, 13.3 vs. 12.0 L/calf/d, SE=0.7; CF vs. NCF). Correspondingly, meal frequency and meal time evolved differently over time ($P<0.03$); calves in CF pens increased meal time and meal frequency, while meal frequency and meal time decreased in NCF pens (wk 2, 5.8 vs. 11.1 meals/d, SE=1.4, and 48.5 vs. 87.1 min meal time/d, SE=15.9, and in wk 4-6, 6.5 vs. 5.1 meals/d, SE=0.5, and 50.8 vs. 38.7 min meal time/d, SE=4.6; CF vs. NCF). Synchronized meal time was greater for calves in NCF pens (24.9 vs. 13.2 min/d, SE=6.1, $P=0.4$), and similar over time ($P=0.2$). However, unsynchronized meal time was subject to a treatment by week interaction ($P=0.02$) corresponding to differences in total meal time, increasing for calves in CF pens between week 2 and 4, and decreasing in NCF pens. Competition at the teat was consistently greater for calves in CF pens compared to NCF pens (during synchronized meal time, 1.2 vs. 0.2 displacements/min, SE=0.3, $P=0.002$). These results indicate that calves are able to adapt to competition for access to teats by adjusting meal patterns, while maintaining time spent engaged in synchronized meals.

The effect of different nutritional regimes on locomotory ability and lying behaviour of replacement gilts

Amy Jean Quinn[1,2], Laura Boyle[2] and Laura Green[1]
[1]University of Warwick, Life Sciences Department, Coventry, CV4 7AL, England, United Kingdom, [2]Teagasc, Pig Development Department, Animal & Grassland Research & Innovation Centre, Moorepark, Fermoy, Co. Cork, Ireland; amy.quinn@teagasc.ie

Diets specifically formulated for replacement gilts (i.e. developer diets) may help to reduce lameness and ultimately improve longevity and productivity. The aim of this study was to compare a developer diet (D) with a gestating sow (G) and finisher (F) diet for replacement gilts in terms of locomotory ability and lying behaviour. One hundred and eighty Large White × Landrace gilts, housed in 18 pens (10 pigs/pen) were selected at ~ 65 kg. Pens were blocked on weight and allocated at random to the following treatments: D (n=6), F (n=6) G (n=6) from 65 kg to 140 kg (c. service weight) over a 12 wk period. Four focal pigs per pen were identified on the basis of weight and lameness score for behaviour measurements. Locomotory ability was scored from 0-5, weekly. Pigs with scores greater than 1 were considered lame. Behaviour of focal pigs was recorded by instantaneous scan sampling at 10 min intervals over 24 hrs on week 1, 3, 6, 9 and 12. Four behaviours were recorded; standing (St), dog sitting (Ds), lying (Ly) and feeding (Fd). Lameness data were analysed by logistic multinomial regression analysis with repeated measures (proc genmod) and behaviour data were analysed using mixed models of analysis of variance for repeated measures (proc mixed), non-parametric data were transformed by square root transformation. Both F and G gilts had a higher risk of lameness (2.22 [Odds ratio=OR]; 1.40-3.50 [CI] and 2.10 [OR]; 1.27-3.34 [CI] respectively; $P<0.01$) than D gilts. There was no effect of treatment and lameness on standing (0.06±0.03 as a mean proportion of 1 per 24 hrs), sitting (0.02±0.02), lying (0.87±0.04) and feeding (0.03±0.02) behaviour ($P>0.05$). Feeding replacement gilts a diet specifically formulated to meet their developmental needs led to improved locomotory ability compared to diets formulated for gestating sows and finisher pigs. However, there was no effect of diet on general patterns of postural behaviour/activity.

Group-housed sows who engage in aggression after mixing have reduced injuries and gain more weight

Megan Verdon[1], Rebecca Morrison[2], Maxine Rice[1] and Paul. H. Hemsworth[1]
[1]Animal Welfare Science Centre, Melbourne School of Land and Environment, University of Melbourne, Parkville, Vic, 3010, Australia, [2]Rivalea Australia, Corowa, NSW, 2646, Australia; meganjverdon@gmail.com

There is some limited evidence in the literature that aggressive behaviour of individual sows affects how well individuals perform in a static group. Thus this study examined the relationships between individual sow aggression and welfare in group-housed sows. 200 parity 1 sows were randomly mixed into pens of 10 (1.8 m²/sow, a space allowance that is internationally common) within 7 days of insemination and floor fed four times per day (07:30, 09:00, 11:00, 15:00 h). Records were taken on aggression (delivered and received) at feeding, fresh skin lesions and plasma cortisol concentrations on day 2 (labelled D2), D9 and D51 after mixing, liveweight at D2 and D100 and litter size (total). Sows were classified at D2 as 'Submissive' if they delivered no aggression, 'Subdominant' if they received more aggression than they delivered and 'Dominant' if they delivered more aggression than they received. Data were appropriately transformed and relationships examined using a one-way ANOVA with classification as the between-subjects factor. Multiple comparisons between means were performed using the LSD test. Submissive sows sustained more injuries on D9 ($F_{2,193}$=10.9, $P<0.001$) and D51 ($F_{2,174}$=6.8, $P<0.05$), but there was no difference between classifications and injuries on D2 ($F_{2,196}$=2.0, $P=0.14$). Dominant sows gained more weight ($F_{2,141}$=5.1, $P=0.01$), tended to have larger litters ($F_{2,164}$=2.1, $P=0.10$), and had the lowest cortisol at D2 ($F_{2,196}$=4.0, $P=0.02$). At mixing, aggressive sows engage in fights for dominance, while others do not deliver aggression. Once a hierarchy is established, dominant sows have reduced risk of receiving aggression, and hence reduced injuries, and have greater weight gain. In conclusion, sows that engage in aggression at mixing and gain dominance have reduced injuries and stress. Their increased weight gain and litter size may be due to increased feed intake through priority access to feed and/or less stress.

The effects of non-uniform environmental conditions on piglet mortality and behavior of sows

Gabriela M. Morello[1], Jeremy Marchant-Forde[1], Donald C. Lay, Jr.[1], Brian T. Richert[1] and Luiz Henrique A. Rodrigues[2]
[1]*Purdue University, Animal Sciences, 125 South Russell Dr., West Lafayette, IN, USA., 47907, Poultry Building, USA,* [2]*Unicamp, Agricultural Engineering, Av. Candido Rondon, 501 Barão Geraldo – Campinas /SP Cidade Universitária Zeferino Vaz, 13083-875, Brazil; gabymm@gmail.com*

Studies have shown a wide variability of total piglet mortality and crushing rates among distinct litters, which has been associated with sow behavior, in terms of occurrence and duration of posture changes. In an effort to understand factors that affect sow maternal ability, this study aimed to evaluate how non-uniform environmental conditions within commercial farrowing rooms affect piglet mortality and sow behavior. Temperature, relative-humidity and light-intensity were recorded every minute in ten different locations within each of two farrowing rooms at the Purdue Animal Sciences Research and Education Center, during November (2012) through January (2013). Environmental data and sow behavior (posture changes) were evaluated for 48 hours post-farrowing. Production data records were retrieved for the months studied, from 2008 through 2012. Total piglet mortality and crushing rates were evaluated as functions of parity, crate location, month of year and total number of piglets born alive, through the GLM procedure on SAS. Piglet mortality was affected by parity ($P=0.06$), month ($P<0.01$), total live born piglets ($P<0.01$) and non-uniform environmental conditions ($P=0.03$). The highest average mortality rate (1.79 ± 0.14 piglets/sow) was obtained during January months. Parity nine sows tended ($P<0.10$) to crush more piglets (1.50 ± 0.51 piglets/sow) than parity 3 or less sows ($<0.61\pm0.10$ piglets/sow). Temperature differences between crates were up to 5 °C (range 19-24 °C), whereas relative-humidity variations were up to 10% (range 35-45%) and average light intensity measurements were up to 250 Lux different (range 50-320 Lux). Although total number of posture changes were not affected by non-uniform environmental conditions, sows under colder conditions within same room took longer (13.17 ± 8.69s, $P=0.05$) to lie laterally from a standing posture than those in warmer conditions (6.96 ± 3.22s). Results indicated a need for a better understanding of the relationship between micro-climates and behavior of sows and piglet mortality.

Indirect genetic effects and housing conditions influence aggressive behaviour in pigs

Irene Camerlink[1,2], Simon P. Turner[3], Piter Bijma[1] and J. Elizabeth Bolhuis[2]
[1]*Wageningen University, Animal Breeding and Genomics Centre, De Elst 1, 6700 AH Wageningen, the Netherlands,* [2]*Wageningen University, Adaptation Physiology Group, De Elst 1, 6700 AH Wageningen, the Netherlands,* [3]*SRUC, Roslin Institute Building, Easter Bush, Midlothian EH25 9RG, Edinburgh, United Kingdom; irene.camerlink@wur.nl*

Indirect Genetic Effects (IGEs), also known as social genetic effects or associative effects, are the heritable effects that an individual has on the phenotype of its social partners. Selection for IGEs has been proposed as a method to reduce harmful behaviours, in particular aggression, in livestock. The mechanisms behind IGEs, however, have rarely been studied. Objective was to assess aggression in pigs which were divergently selected for IGEs estimated on growth. In a one generation selection experiment, we studied 480 offspring of pigs (*Sus scrofa*) that were selected for high or low IGE. Pigs were, from 5-23w of age, housed by separate IGE classification in barren or straw-enriched pens. Aggressive behaviours and skin lesion scores, a proxy measure of aggression, were recorded and analysed on pen level in a mixed model. Results are means ± SEM in percentage of behavioural scans. High IGE pigs, selected for a positive effect on growth of group members, performed less biting at three days after weaning (High 0.11 ± 0.01; Low 0.17 ± 0.01; $P<0.01$) and a week after a 24h regrouping test (High 0.02 ± 0.01; Low 0.07 ± 0.01; $P<0.01$). Remarkably, high IGE pigs showed half as much aggression at reunion with familiar group members after they had been separated during the regrouping test (High 7 ± 1.6; Low 15 ± 3.0; $P<0.001$). The IGE groups did not differ in amount of fighting or in number of skin lesions, both under stable social conditions and in confrontation with unfamiliar pigs. The straw-enriched environment was associated with more skin lesions ($P<0.05$), but less aggression under stable social conditions ($P<0.05$). Changes in aggression between pigs selected for IGE were not influenced by G×E interactions with regard to environmental enrichment. In conclusion, selection on IGE can potentially reduce aggression in pigs but more likely targets a behavioural strategy, rather than a single behavioural trait such as aggressiveness.

Chopped or long roughage: what do calves prefer?

Laura Webb[1], Margit Bak Jensen[2], Bas Engel[3], Kees Van Reenen[4], Imke De Boer[1], Walter Gerrits[5] and Eddie Bokkers[1]
[1]*Animal Production Systems, Wageningen University, P.O. Box 338, 6700 AH Wageningen, the Netherlands,* [2]*Dept. of Animal Science, Aarhus University, P.O. Box 50, 8830 Tjele, Denmark,* [3]*Biometris, Wageningen University, P.O. Box 100, 6700 AC Wageningen, the Netherlands,* [4]*Livestock Research, Wageningen University, P.O. Box 65, 8200 AB Lelystad, the Netherlands,* [5]*Animal Nutrition, Wageningen University, P.O. Box 338 Wageningen, 6700 AH Wageningen, the Netherlands; laura.webb@wur.nl*

Longer roughage particles increase rumination and may improve the welfare of calves fed little roughage. Double demand function quantifies the strength of the relative preference for substitutable resources. Animals work for two resources simultaneously, with the price of the resources changing relative to each other. With this method, the preference for two particle lengths (chopped versus long) of two roughage types (hay and straw) was quantified in nine male calves fed milk replacer and concentrates from 10 l/d and 0.3 kg/d at 2 months, to 16 l/d and 2.7 kg/d at 6 months of age. Calves were habituated to each roughage type, and then trained to press two panels for a 10 g lucerne reward. When calves consistently accessed more rewards on the panel with the lowest price (less presses), they were tested. The pairs of prices (i.e. number of presses required for a reward) for simultaneously available chopped and long roughage rewards were (chopped/long): 7/35, 14/28, 21/21, 28/14, 35/7. Demand functions were estimated per roughage type and per calf. The cross-points of demand functions for chopped and long roughage were estimated for hay and straw. The deviation of the cross-point from the midpoint (here 21) indicates the strength of the preference. Calves showed a preference for long compared with chopped hay ($P=0.002$) with a deviation from the midpoint of 5.4±0.9. No preference was found for chopped versus long straw ($P=0.711$), with a deviation of 1.9±3.0. The large variation for the latter comparison is mostly due to 3 calves showing a preference for one of the straw types. Calves worked consistently for access to roughage, and exhibited a preference for long hay over chopped hay. The lack of effect for straw may be because straw is coarse, providing enough structure regardless of particle length.

Dairy cattle preferences for sprinklers delivering different water volumes

Jennifer M. Chen[1], Karin E. Schütz[2] and Cassandra B. Tucker[1]
[1]UC Davis, Davis, CA, USA, [2]AgResearch, Ltd, Hamilton, New Zealand; jmchen@ucdavis.edu

Sprinklers effectively reduce heat load in cattle, but elicit variable behavioral responses: in some studies, cattle readily use spray, but avoid it or show no preference in others. Some of this behavioral variation may be explained by spray attributes that differ across studies. Our objective was to examine preferences for sprinklers with different water volumes and droplet sizes in a shaded Y-maze. Lactating cows (n=18) were tested in summer (testing air temperature 29±5.1 °C, mean ± SD) in 3 pairwise comparisons between either a no-spray Control, a sprinkler delivering 0.4 l/min with smaller droplets (Low), and/or a 4.7 l/min sprinkler with larger droplets (High). For each pairing, cows were exposed to each treatment 5 times before allowing them to choose between treatments once daily (12 min/d) for 8 consecutive d. Cattle received no other water cooling 3 h before and after testing. In all 3 comparisons, when cows chose the treatment delivering more water, respiration rate was lowered over the 12-min test ($P \leq 0.058$, GLMM). Although sprinklers reduced respiration rate, cows tended to choose Low over Control only 69% of the time (SE: 9.3%; $P=0.096$, Wilcoxon signed-rank test), and showed no overall preferences in the other comparisons (preference for High: 58% vs. Control, SE: 9.4%; 42% vs. Low, SE: 9.6%; $P \geq 0.552$). However, when choosing between Control and sprinklers, body temperature affected preference: at normal temperature (≤ 38.6 °C), cows chose High 51% and Low 67% of the time, but at ≥ 39.3 °C they chose High 72% and Low 76% of the time (each 1 °C rise in body temperature increased the odds of choosing High and Low 3 and 1.4 times, respectively; GLMM). In conclusion, cattle did not show clear preferences based on spray attributes, but body temperature influenced choice.

Housing type affects the lying behaviour of dairy cows given access to an outside 'loafing paddock'

Fritha Langford[1], David Bell[1], Daniel Chumia[1], Rhonda-Lynn Graham[1], Mhairi Jack[1], Ian Nevison[2], Dave Roberts[1] and Marie Haskell[1]
[1]*SRUC (Scotland's Rural College), Animal and Veterinary Sciences, King's Buildings, West Mains Road, Edinburgh, EH9 3JG, United Kingdom,* [2]*Biomathematics and Statistics Scotland (BioSS), James Clerk Maxwell Building, The King's Buildings, Mayfield Road, Edinburgh, EH9 3JZ, United Kingdom; fritha.langford@sruc.ac.uk*

It has been shown that cows housed in cubicles are motivated to use fields for lying. It is not known whether improving lying comfort within the house affects this choice. Four groups of 12 lactating dairy cows were used to investigate housing-type effects on preferences for an outside loafing-paddock (OLP). The cows were milked 3×/d and bore 'IceTags' to record standing and lying over 24h/d. There were two housing treatments: cantilever cubicle housing with rubber-matting and liberal sawdust-bedding (CUBS) and straw-yard loose-housing (STRAW). Each group of cows were housed in each treatment condition. Habituation to the experimental housing for 1 d was followed by 5 d of behavioural observations inside the housing without access to the OLP ('Baseline'). Cows were trained for 1 d to access the OLP. This was followed by 5d of behavioural observations with free access to the housing area and the OLP from post feed-delivery until milking and again until 19:30 ('Preference'). Every 20 mins instantaneous scans of the cows were undertaken with place, behaviour, spatial information and two nearest neighbours were recorded for each cow. When cows were lying down, exact lying posture was recorded. During the 'Preference' period, cows in CUBS spent more time in the OLP than cows in STRAW (79.8±0.02% vs 50.8±0.09%, GLMM, Wald=4.8, $P<0.05$). They also spent a higher proportion of the time lying down in the OLP than when housed in STRAW (56.7±0.03% vs 18.3±0.06%, GLMM, Wald=6.4, $P<0.01$). In the OLP, cows in CUBS spent 6.4±0.02% of the lying time lying laterally. Cows in STRAW showed lateral-lying in the house (3.2±0.01%) and in the OLP (1.2±0.01%). Cows in CUBS used the OLP primarily for lying. These results have implications for the comfort achievable in cubicle housing, in many designs of which it is difficult for cows to lie laterally.

Assessing motivation to access a food reward in laying hens while kept in a social group

Carlos E. Hernandez[1,2], Caroline Lee[1], Jim Lea[1], Drewe Ferguson[1], Sue Belson[1] and Geoff Hinch[2]
[1]CSIRO, Animal, Food and Health Sciences, New England Highway, 2350 Armidale, NSW, Australia, [2]University of New England, School of Environmental and Rural Science, University of New England, 2351 Armidale, NSW, Australia; carlos.hernandez@une.edu.au

The aim of this study was to validate a methodology to assess individual motivation to access a resource while within a social group. We hypothesised that individual motivation would be minimally influenced by conspecifics with contrasting motivation levels when tested as a group. Eighteen ISA brown hens (fitted with RFID leg bands) were trained to jump over a platform of adjustable height (4 available) to access a food reward (mixed grains). During training, the maximum price paid (MPP- maximum height jumped) to access the food after a 5 h food restriction was recorded for all hens (trained as one group) as a basal measure of motivation to access the resource (MPP-B). Based on this measure, hens were randomised equally to either an L group (n=9): *ad libitum* mixed grains 30 min before the test or an H group (n=9): food restricted for 5h before the test. The groups were then combined and tested together with the cost to access the food increasing in 20cm increments from 30 cm to 70 cm and in 10 cm increments thereafter (1 d per cost), until all hens stopped responding. Least squares means analysis showed that MPP-B was similar for both groups. The MPP during the test was lower in the L group than in the H group (L=76.2±5.4 vs H=92.7±5.4, $P{\leq}0.05$) and initial MPP-B did not have a significant effect on MPP during testing. In conclusion, feeding level prior to testing was shown to significantly influence motivation in hens even though both groups were tested together. These results suggest that individual motivation for a food resource is not significantly altered by social facilitation and therefore can be assessed without individual isolation. Therefore this methodology may be relevant in evaluating motivation in large groups rather than using traditional behavioural demand methodologies requiring animals to be tested in isolation.

Emotional and behavioural responses in gun dogs during blind-retrieving tasks in a T-maze

Carla Torres-Pereira and Donald M. Broom
Centre for Animal Welfare and Anthrozoology, Department of Veterinary Medicine, University of Cambridge, Madingley Road, Cambridge CB3 0ES, United Kingdom; dmb16@cam.ac.uk

Are gun dogs' behavioural responses during retrieving affected by a reprimand and by the nature of the objects to be retrieved? Are there differences in heart-rate before and after dogs do something wrong? Finally, are there behavioural and emotional differences when a dog does something wrong and when the same dog does what it was supposed to, in the absence of a reprimand? A combination of field trials with blind-retrieving tasks was designed to characterise behavioural and emotional responses in 32 gun dogs during the retrieval of different objects hidden at the top extremities of a T-maze (right: canvas dummy, first retrieval; left: rabbit skin dummy or dead pigeon, second retrieval) delineated inside a fenced field with different treatments from the owner, who could either reprimand the dogs or only give directions. Data from video recording and Polar heart-rate monitor recording were analysed with GenStat. The dogs stopped and looked at the owner more often when there was a reprimand (chi-square test: $\chi^2=50.14$, df=1, $P<0.001$). This behaviour depended on the object to be retrieved, being more frequent for the canvas dummy (chi-square test: $\chi^2=13.24$, df=1, $P<0.001$), the retrieval of which originated more off-track (chi-square test: $\chi^2=14.76$, df=1, $P<0.001$) and off-field (chi-square test: $\chi^2=24.47$, df=1, $P<0.001$) faults. After doing wrong, the dogs' heart-rates tended to be higher for the second retrieval (Wilcoxon signed-rank test: T=2347, n=108, $P=0.056$) and were higher specifically for the dead pigeon's retrieval (paired t-test: t=2.48, df=57, $P=0.016$). In the absence of a reprimand, the dogs stopped and looked at the owner more often when they did something wrong (chi-square test: $\chi^2=12.60$, df=1, $P<0.001$) and their heart-rates were lower after doing right (paired t-test: t=-2.20, df=27, $P=0.037$). Dog's behavioural and emotional responses during retrieving are affected by the object to be retrieved, own action and owner's behaviour.

Dog owners' perception of their dogs' responses to loss of a canine companion

Leticia Fanucchi[1,2], Joyce Mcpherrin[1] and Ruth C. Newberry[1,2]
[1]Washington State University, Center for the Study of Animal Well-being, Department of Veterinary and Comparative Anatomy, Pharmacology and Physiology, Pullman, WA 99164-6520, USA, [2]Washington State University, Department of Animal Sciences, P.O. Box 646351, Pullman, WA 99164-6351, USA; l.fanucchi@email.wsu.edu

Grief refers to the negative feelings generated by the long-term loss of a significant social companion, usually due to death. Grief is a concept typically reserved for humans, and separation distress symptoms are characteristic of grief in humans. However, separation distress symptoms have been observed in non-human animals experiencing social loss. Separation anxiety is well documented in domestic dogs, with characteristic behaviors displaying emotional distress. We hypothesized that an adult dog socially attached to another dog living in the same household, when experiencing the loss of the other dog, would exhibit grief-like behavioral changes following the loss. An online survey was conducted of dog owners in the United States. The questionnaire comprised questions assessing what the owners perceived to be their dog's reactions in the month after vs. the week before the loss of a canine companion. Wilcoxon Signed-rank test statistics showed that significant behavioral changes reported by owners (n=28) included decreased appetite ($P=0.04$), depressed posture ($P=0.002$), decreased play activity ($P=0.001$), and increased restlessness ($P=0.003$). A parallel control survey indicated no owner-reported change in appetite, posture or restlessness over an equivalent period when a dog with a social attachment to another dog in the same household experienced no loss of the canine companion (n=117, $P>0.05$). Owners reported an increase, rather than a decline, in play ($P<0.001$) over the same period. These data support the hypothesis that domestic dogs experience grief during the month following the loss of a canine companion, although direct observation will be required to establish whether owner perception accurately reflects actual behavioral changes. By understanding the relationships formed by domestic dogs with their conspecifics, this knowledge can potentially offer insights to improve companion animals' wellbeing.

Welfare of community dogs in Campo Largo-PR/Brazil: preliminary results

Larissa Helena Ersching Rüncos[1], Gisele Sprea[2] and Carla Forte Maiolino Molento[1]
[1]Universidade Federal do Paraná/UFPR, Laboratório de Bem-estar Animal/LABEA, Rua dos Funcionários, 1540, 80035050, Brazil, [2]Prefeitura Municipal de Campo Largo, Setor de Controle de Zoonoses e Bem-estar Animal, Av. Padre Natal Pigatto, 925, 83601630, Brazil; lari.hr@gmail.com

Community dogs are stray dogs registered and spayed/neutered by the municipality, monitored and fed daily by registered caretakers from the community and provided with shelter and medical care. Our objective was to evaluate the welfare of dogs registered in the community dog program in Campo Largo, Brazil. The welfare of 53 dogs was assessed *in loco* through physical exam, analysis of environment, 30 min evaluation of dog responses to stimulus, and interview with caretakers, which together were used to categorize the five freedoms. Specifically regarding freedom from fear and distress, dog attitudes towards caretaker and other individuals were taken into consideration. Freedoms were categorized as respected (RE), moderately restricted (MR) or severely restricted (SR). The welfare status was classified as very high, high, regular, low or very low, based on the most frequent category per freedom and intermediate possibilities. Thirty-five (66.1%) dogs were females and 18 (33.9%) males, with ages varying from approximately 0.5 to 15 years-old. Freedom to express normal behavior was considered RE for all dogs. Freedom from hunger and thirst was considered RE for 19 (35.8%) dogs, MR for 31 (58.5%) and SR for 3 (5.6%) SR. Freedom from discomfort was considered RE for 13 (24.5%) dogs, MR for 29 (54.7%) and SR for 11 (20.7%). Freedom from pain, injury and disease was considered RE for 7 (13.2%), MR for 40 (75.5%) and SR for 6 (11.3%). Freedom from fear and distress was considered RE for 4 (7.5%) dogs, MR for 42 (79.2%) and SR for 7 (15.1%). The welfare of the community dogs from Campo Largo was very high for 4 (7.5%) dogs, high for 30 (56.6%), regular for 12 (22.6%) and low for 7 (13.1%). Results show that for most of the studied dogs, welfare was good. However, important welfare restrictions exist and need improvement.

Do cats' initial responses to a confinement housing environment persist upon re-exposure?

Judith Stella[1], Candace Croney[2] and Tony Buffington[3]
[1]*The Ohio State University, Veterinary Preventive Medicine, 1920 Coffey Rd, Columbus OH 43210, USA,* [2]*Purdue University, Animal Sciences, 125 S. Russell St., W. Lafayette, IN 47907-2042, USA,* [3]*The Ohio State University, Veterinary Clinical Sciences, 601 Vernon L Tharp St, Columbus Oh 43210, USA; stella.7@osu.edu*

The aim of this study was to compare behavioral responses of cats to a controlled confinement experience to their responses one year later. To mimic the experience of cats surrendered to a shelter or admitted to a veterinary hospital, adult cats (n=25) were recruited from university employees, caged singly in a vivarium at the Ohio State University and randomly assigned to one of two treatment groups. M+ was an enriched macro (room)-environment (minimal noise, disruption, consistent schedule); M- was an unenriched macro (room)-environment (recorded dog barks, frequent disruptions, unpredictable schedule). Cats were housed for 2 days, observed from 08:00-16:00 for maintenance/affiliative behaviors (e.g. eating, drinking, elimination, soliciting caretaker interaction) and agonistic/ avoidant behaviors (e.g. growling, hissing, hiding) using scan sampling (every 2 hours) and 5 minute focal sampling (alternate hours), then returned to their owners. The study was repeated one year later and results were compared. Data analysis by GLMM revealed that more cats housed in the M+ environment ate on day one and day two in year one and year two than cats housed in the M- environment (treatment $P=0.007$, day $P=0.004$). Pearson's chi^2 test of scan samples of the number of cats exhibiting affiliative/maintenance behaviors in treatment M+ and M- in year one and year two revealed statistical significance at time points 1 ($P<0.0001$), 2 ($P=0.004$), 4 ($P<0.0001$), 5 ($P<0.0001$), 6 ($P<0.0001$), 8 ($P=0.001$). Wilcoxon sign rank test of the average number of cats hiding revealed a significant increase from year one to year two on day one for cats housed in M- ($P=0.03$) but not for cats housed in M+. These results suggest that cats may maintain a memory of previous confinement experiences and that enriched housing is important to cat welfare both during initial exposure and to facilitate adaption upon re-exposure in veterinary hospitals and shelters.

Is the Strange Situation Procedure a reliable method to use when investigating attachment in dogs?

Therese Rehn, Ragen T.S. Mcgowan and Linda Keeling
Swedish University of Agricultural Sciences, Department of Animal Environment and Health, P.O. Box 7068, 750 07 Uppsala, Sweden; Therese.Rehn@slu.se

The Strange Situation Procedure (SSP) originally developed for children is increasingly being used to study attachment between dogs and humans. In this experiment, 12 female beagle dogs were tested in two treatments to identify possible order effects in the test, a potential weakness in the SSP. In one treatment (FS), dogs participated with a 'familiar person' and a 'stranger'. In a control treatment (SS), the same dogs participated with two unfamiliar people, 'stranger A' and 'stranger B'. Comparisons were made between episodes within as well as between treatments. As predicted in FS, dogs explored more (T=39, $P<0.001$) in the presence of the familiar person (0.20 ± 0.03 (mean ± SE)) than the stranger (0.10 ± 0.02). Importantly, they also explored more (T=26.5, $P=0.004$) in the presence of stranger A (who appeared in the same order as the familiar person and followed the same procedure) (0.16 ± 0.03) than stranger B (0.07 ± 0.02) in SS. Furthermore, there was no difference between treatments in how much the dogs explored in the presence of the familiar person in FS and stranger A in SS. In combination, these results indicate that the effect of a familiar person on dogs' exploratory behaviour, a key feature when assessing secure attachment styles, could not be tested reliably due to order effects in the SSP test. It is proposed that in the future only counterbalanced versions of the test are used. It is further proposed that an alternative approach might be to compare the response of the dog at reunion, since it was found that dogs reliably initiated more contact with the familiar person compared to the strangers. This could be achieved by focussing either on the behaviour of the dog in those episodes of the SSP when the person returns, or on reunion behaviour in other studies.

Development and validation of a behavioural test to discriminate between bold and shy cats

Jacklyn J Ellis, Vasiliki Protopapadaki, Henrik Stryhn, Jonathan Spears and Michael S Cockram
Sir James Dunn Animal Welfare Centre, Department of Health Management, Atlantic Veterinary
College, University of Prince Edward Island, 550 University Ave, Charlottetown, PE, C1A 4P3,
Canada; jjellis@upei.ca

Whether a cat is bold or shy could potentially affect its response to a shelter environment and the types of behaviours exhibited to potential adopters. A simple test to discriminate between the two could help shelters allocate relevant types of environmental enrichment to improve cat welfare. To develop such a test, nine cats were subjected to five variations per week of approach or novel object tests in an open-field, for up to five weeks. In a GLM, measures that differed between cats ($P<0.05$) but were not affected by time ($P>0.05$), were entered into a principal component analysis. Although sample size was small, the bold/shy component described the most variance (37%). Additionally, observer ratings for this behavioural style had high inter-observer agreement (concordance correlation coefficient =0.83, κ=0.78) and categorised individuals identically to the bold/shy component. As determined by component loadings and practicality, latency to emerge from and percentage of time spent in a carrier showed the most promise to discriminate between bold/shy cats. Subsequently, these tests were investigated every other day, for up to nine days, using seventeen different cats. In a GLM, these measures remained consistent over time ($P>0.05$), and different between cats ($P<0.05$). Observer ratings showed substantial inter-observer agreement when dichotomising bold/shy individuals (κ=0.62). For both tests, data from the two studies were analysed to identify the cut-point with the highest percentage of correct classification (PCC) overall, as well as for bold/shy cats separately. Latency to emerge from the carrier with a cut-point of 10 seconds was deemed the most appropriate test for discriminating between bold/shy cats because it had a high PPC overall (81), within bold (74) and shy (100) cats respectively, was quick and easy to administer, and was best suited to identify correctly shy individuals, who arguably could derive greatest benefit from identification and extra attention.

Can pigmentation and hair cortisol predict social strategies in goats?

Judit Vas, Rachel Chojnacki and Inger Lise Andersen
Norwegian University of Life Sciences, Department of Aquaculture and Animal Sciences, P.O.
Box 5003, 1432, Aas, Norway; judit_banfine.vas@umb.no

We addressed the validity of hair cortisol as an indicator of stress during pregnancy and its relationship to social strategies and blood cortisol in goats. Fifty-four Norwegian dairy goats were used (18 colored and 36 white animals, 3.0±0.2 and 2.7±0.1 years, respectively). Initially, vocalizations when separated from group mates and reaction to an approaching human (boldness) were observed. Afterwards, during early pregnancy, they were put into groups of 1, 2 or 3m^2 per animal (colored and white goats balanced) and kept in these groups until kidding. The hair grown during the treatment period (14 weeks) in a 20 cm^2 area from the front region of neck was analyzed. In the first, second and last third of pregnancy blood samples were taken (three occasions, two consecutive days) and social behaviors of animals were observed (9 hours). We tested the effect of treatment and pigmentation on hair cortisol levels and social behaviors (Genmod procedure), the relationship between blood cortisol, hair cortisol and behavioral variables (Spearman rank correlation), and compared vocalization and boldness in pigmented and white goats (Wilcoxon signed rank tests). Higher hair cortisol levels were connected to white color ($P=0.006$, white: 6.0±0.6 mg/g, colored: 4.6±0.3 mg/g), more frequent vocalizations when separated ($P=0.008$) and more offensive behaviors shown ($P=0.016$). Cortisol levels in blood, boldness, density treatment, socio-positive behavior and defensive behaviors did not have a relationship with cortisol levels in hair. Colored goats showed bolder ($P=0.043$) and more defensive behaviors ($P=0.034$); however, pigmentation did not have an effect on blood cortisol level, vocalization or the other social behaviors. In sum, a higher level of hair cortisol appears to be connected to white pigmentation, active, anxiety-like behaviors during separation and an active social strategy in the group. Hair cortisol levels can be a good indicator of inter-individual differences.

Reducing aggressive behaviour in young piglets by cognitive environmental enrichment

Lilia Thays Sonoda[1], Michaela Fels[1], Sally Rauterberg Rauterberg[1], Maciej Oczak[2], Gunel Ismayilova[3], Stefano Viazzi[4], Marcella Guarino[3], Erik Vranken[2], Daniel Berckmans[4] and Jörg Hartung[1]
[1]University of Veterinary Medicine Hannover, Foundation, Institute for Animal Hygiene, Animal Welfare and Farm Animal Behaviour, Bünteweg 17p, 30559 Hannover, Germany, [2]Fancom Research, Industrieterrein 34, 5981 NK Panningen, the Netherlands, [3]Faculty of Veterinary Science, Department of Veterinary Science and Technologies for Food Safety, Via Celoria, 10, 20133 Milano, Italy, [4]M3-BIORES: Measure, Model & Manage Bioresponses, M3-BIORES: Measure, Model & Manage Bioresponses, Kasteelpark Arenberg 30, 3001 Leuven, Belgium; lilia.thays.sonoda@tiho-hannover.de

It is known that pigs raised in enriched environments express less abnormal and aggressive behaviour than pigs housed in barren pens. A new method of cognitive environmental enrichment was experimented at the research farm of University of Veterinary Medicine in Hannover, Germany. In the first study phase, 8 entire litters of 8-12 suckling piglets, 25 days old with an average weight of 7 kg, were trained in their farrowing pens where they had to learn the link between a sound given by an electronic feeder and a feed reward in form of chocolate candies during a period of 8 days. In the second study phase, all trained piglets were used in pairs, as resident and as intruder, in a resident-intruder test, on the 4th day after weaning, aiming to study the potential of the feeding system to break aggressive behaviour. For training data, ANOVA and posthoc tests (SNK) were conducted. Resident-intruder data were analysed by chi square test. The analysis of training revealed that piglets learned the link between sound and feed during 8 days and the number of piglets around the feeder awaiting chocolate candies after sound increased with consecutive training days ($P<0.05$). In the resident-intruder test, 260 aggressive interactions were analysed and on average 80% of aggressive interactions were broken by activation of the feeder when fighting started ($P<0.05$). Also fighting that restarted was interrupted. In 55% of all fights stopped by feeder activation, the aggressor reacted to the feeder. If we considered only fights where the aggressor reacted to the feeder, slightly more fights could be stopped (97% vs 93%). We conclude that the electronic feeding system has the potential to be used as cognitive enrichment for suckling and weaned piglets being able to reduce aggressive behaviour in weaned piglets.

Inter-logger variation of spatial proximity devices: consequences for animal social networks

Natasha Boyland[1], Richard James[2], David Mlynski[3], Joah Madden[1] and Darren Croft[1]
[1]University of Exeter, Psychology, Centre for Research in Animal Behaviour, College of Life and Environmental Sciences, Exeter, Devon. EX4 4QG, United Kingdom, [2]University of Bath, Physics, Centre for Networks and Collective Behaviour, University of Bath, Bath, BA2 7AY, United Kingdom, [3]University of Bath, Biology and Biochemistry, Centre for Networks and Collective Behaviour, University of Bath, Bath, BA2 7AY, United Kingdom; nkb204@exeter.ac.uk

Social network analysis has become an increasingly popular method to link individual behaviour to population level patterns (and vice versa). Technological advances of recent years, such as the development of spatial proximity loggers, have enhanced our abilities to record contact patterns between animals. However, loggers are often deployed without calibration which may lead to sampling biases and spurious results. In particular, loggers may differ in their performance (i.e. some loggers may over-sample and other loggers under-sample social associations). However, the consequences of inter-logger variation in performance has not been thoroughly considered or quantified. In this study, 20 proximity loggers made by Sirtrack Ltd. were fitted to 20 dairy cows over a three week period. Contact records resulting from field deployment demonstrated variability in logger performance when recording contact duration. A value of comparative performance, which can be considered the 'bias in detecting other loggers', was calculated and ranged from -14.8% to 22.1% across the sample, and was highly consistent for each logger over time (r=0.992, 95%CI 0.982-0.996). Testing contact initiation distances of loggers under standardised conditions further revealed consistent differences between loggers, which were significantly positively correlated with the aforementioned logging bias (r=0.457, n=20, P=0.043). This suggests that the inter-logger variation observed during field deployment was partly due to intrinsic variation in devices. We demonstrate potential consequences of this variation in logging performance, for social network analysis; particularly how measures of connectivity can be influenced by logging performance. Finally, we suggest some approaches to correct data generated by proximity loggers with varied performance, that could be used to improve the robustness of future analysis.

Association of social rank with health, reproduction, and milk production of dairy cows

Marcia Endres, Karen Lobeck, Paula Basso Da Silva and Ricardo Chebel
University of Minnesota, 1364 Eckles Avenue, St. Paul, MN 55108, USA; miendres@umn.edu

The objective of this study was to investigate the association of social rank during the close-up prepartum period with health, reproduction, and milk production of dairy cows during early lactation. The study was conducted in a freestall sand-bedded dairy farm in south-central Minnesota. One hundred and ninety Jersey cows were enrolled in the study 5 weeks prior to expected calving date and housed in pens of 44 cows with a enrolling stocking density of 90% of headlocks and 100% of stalls. Cows were balanced for body condition score. Cows with locomotion score >2 (in a 1 to 5 scale, where 3=lame) were not included in the study. Displacements from the feed bunk were measured for 3 hours following fresh feed delivery for 4 days each week. Most cows were observed for 4 weeks. A displacement index was calculated as the number of displacements initiated by a cow divided by the number of displacements initiated plus number of displacements received by a cow. Cows with an average displacement index of <0.4 were categorized as low-ranking, 0.4 to 0.6 middle-ranking, and >0.6 were considered high-ranking. Each cow received an average displacement index (each day the cow received an index score). There were 89 low, 59 middle and 42 high ranking cows in the study. Health events for the first 100 DIM, milk production and composition for the first 3 monthly milk tests, and first breeding pregnancy rate were recorded for each cow. The Logistic procedure was used to evaluate health and reproductive events. The Mixed procedure was used to analyze milk production and milk composition. There was no association of social rank with retained placenta, metritis, death, displaced abomasum, and mastitis events. Displacement index was associated with first breeding pregnancy rate ($P<0.01$). Middle rank cows were 3 times more likely to become pregnant after first AI than low ranking cows with no differences between low and high ranking cows. One suggested reason for this could be that the middle ranking cows spent less time and energy establishing hierarchy and that could have helped them transition better into the new lactation. There was no association between milk production and social rank. Percent milk fat from the second test was associated with social rank ($P=0.04$) and it was greater in low ranking cows than high-ranking cows (4.1 ± 0.13 vs $3.7\pm0.16\%$). Middle-ranking cows, however, had similar percent milk fat to low and high-ranking cows. In summary, social rank in the prepartum period was associated with pregnancy status and 2nd test milk fat percentage in early lactation dairy cows. However, pregnancy status may be more complex than simply being associated with social rank. Follow up studies with larger number of animals are warranted.

Comprehension of human social cues in young domestic pigs

Christian Nawroth[1], Mirjam Ebersbach[2] and Eberhard Von Borell[1]
[1]Institute of Agricultural and Nutritional Sciences, Martin-Luther-University, Department of Animal Husbandry & Ecology, Theodor-Lieser-Str. 11, 06120 Halle, Germany, [2]Institute of Psychology, University of Kassel, Department of Developmental Psychology, Holländische Str. 36-38, 34127 Kassel, Germany; nawroth.christian@gmail.com

The emotional reactivity hypothesis claims that domesticated species should perform better in utilizing human social cues than their non-domesticated relatives. So far, only ambiguous results are available for pigs. In this study, we investigated the use of human social cues in an object choice task by 17 domestic pigs (7 wks of age) during the course of four experiments. Subjects had to choose between two bowls of which only one was baited. The baited one was indicated by means of different social cues of the experimenter. In Experiment 1, subjects were able to use proximal momentary and proximal dynamic-sustained pointing cues (n=17; $P=0.002$ and $P<0.001$, respectively) from the start, but failed to use distal cues when the experimenter was in a standing position (all $P>0.05$). However, subjects were able to utilize distal momentary and distal dynamic-sustained pointing cues when cues were provided in a kneeling position (Experiment 2; n= 15; $P=0.001$ and $P<0.001$, respectively). In Experiment 3, subjects chose successful when the experimenter was positioned behind the correct bowl (n=14; $P<0.001$). Nonetheless, when the experimenter, positioned behind the incorrect bowl, pointed towards the correct one, subjects as a group performed on chance level (n=14; $P>0.05$). However, one individual performed significantly above chance, suggesting that local enhancement alone may not explain subjects' performance. Finally in Experiment 4, we found that pigs were also able to use body and head orientation of a human experimenter to locate the baited bowl (n=13; $P=0.001$ and $P=0.015$, respectively). We assume that pigs, even at a very young age, are skillful in utilizing human social cues including the body and head orientation of humans.

How do pigs vocally communicate: a graded or continuous system?

Céline Tallet[1,2], Pavel Linhart[1] and Marek Spinka[1]
[1]*Institute of Animal Science, Ethology department, Přátelství 815, 104 00 Praha Uhříněves, Czech Republic,* [2]*INRA-Agrocampus Ouest, UMR1348 PEGASE, domaine de la prise, 35590 Saint Gilles, France; celine.tallet@rennes.inra.fr*

Vocal signals of pigs are usually described in terms of call categories (e.g. screams, grunts ...). Nevertheless the use of mixed call types suggests that their structure may vary continuously across a quantitative acoustic spectrum. Vocal repertoire of 84 domestic piglets (*Sus scrofa*) in their second week of life was quantitatively described based on 1513 calls recorded in 11 situations: 4 life threatening (e.g. castration), 3 nursing (e.g. just after nursing), and 4 general social (e.g. huddling) situations. We described the acoustic quality of calls with 8 acoustic parameters and did not take into account the piglet identity because we were looking for call types (classical validated method). Based on these parameters, the k-means clustering method showed that the vocal expressions form an acoustic continuum yet with a possibility to distinguish either two or five salient clusters. The analysis of between-situation differences in call expressions revealed that the type of situation is clearly encoded both in the quantitative acoustic properties of the calls, and in the qualitative types of calls used. There was a significant effect of the situation on the 8 acoustic parameters (MANOVA: $F_{10,742}$ = 12.8, $P<0.001$). The distribution of call types among situations was highly non-random (Chi-square from Monte Carlo permutation test, df=10, 2-cluster solution: χ^2=649, $P<0.001$; 5-cluster solution: χ^2=1296.5, $P<0.001$). Dendrograms revealed that in situations of the same category, a similar mixture of call types was used by the piglets. These results suggest that the repertoire of domestic pigs is rather a continuum. However the description as a limited number of call types without losing much of the information tends to show that a categorical perception of conspecific calls might be an efficient cognitive decoding strategy. The type of situation has a direct effect on the vocal emissions, confirming that they are an effective way to transmit the emotional state of the sender.

Dairy calves show negative judgement bias following hot-iron disbudding

Heather W. Neave, Marina A.G. Von Keyserlingk and Daniel M. Weary
University of British Columbia, Animal Welfare Program, 2357 Main Mall, V6T 1Z4, Canada;
hwneave@gmail.com

Cognitive bias tasks involve interpretation of ambiguous stimuli. Depressed or anxious humans are known to judge ambiguous events negatively, and recent work on animals suggests that judgement bias can also provide a method of assessing emotions in non-human animals. Pain is likely the most studied emotion in animals, but most pain research has focused on sensory aspects leaving the emotional experience of pain under explored. Here we present evidence of cognitive bias in response to pain. Dairy calves (n=8) were trained to respond differentially to red and white video screens and then tested with unreinforced ambiguous probe colours (25%, 50% and 75% red) the day before and after the routine practice of hot-iron disbudding, a procedure that is known to be painful for at least 24 hrs and is commonly performed without the use of pain medication. One horn bud was removed on one day (Dehorn 1) and the second bud 4 d later (Dehorn 2). Responses to the probe screens during repeated baseline testing (Mixed model ANOVA, $P>0.05$) suggested the calves did not learn to discriminate the non-reinforced probes. Moreover, responses to the probes after disbudding did not differ between Dehorn 1 and Dehorn 2, so responses from the two sessions were pooled. Analysis of this pooled data showed that calves approached ambiguous screens less frequently after disbudding (73±6, 30±5, 8±5% for the near-positive, halfway, and near-negative probe, respectively) compared to before disbudding (83±3, 33±6, 11±4%, respectively), a difference that was most pronounced for the near-positive probe (Mixed model ANOVA, $P=0.05$). These results indicate calves experiencing pain during the hours after hot-iron disbudding exhibit a 'pessimistic' bias.

Individual housing impairs reversal learning in dairy calves

Rolnei R Daros[1], Joao H C Costa[2], Marina A G Von Keyserlingk[2], Maria J Hotzel[1] and Daniel M Weary[2]
[1]Universidade Federal de Santa Catarina, LETA – laboratorio de etologia e bem estar aniamal, Rodovia Admar Gonzaga, 1346, Itacorubi, Florianópolis, SC, 88034-000, Brazil, [2]University of British Columbia, Animal Welfare Program, Faculty of Land and Food Systems, 2357 Main mall, Vancouver, BC, V6T 1Z4, Canada; rrdaros@gmail.com

There is a growing body of literature suggesting that social housing improves cognitive development in animals. The common practice on dairy farms is to house pre-weaned calves in individual pens but no work to date has tested if this practice impairs their cognitive performance. Holstein calves were either reared individually (n=7) or in a group (n=8) with other calves and cows. Calves were trained twice daily starting at 5 d of age to discriminate between two coloured screens (red and white) using a go/no-go task. 'Go' responses (touching the screen) in response to the positive colour were rewarded with milk. Screen touches following the negative colour were punished with a noise and a 1-min timeout. The discrimination phase started once calves had successfully learned to respond to the positive stimuli (over 90% of correct responses – 20 positive stimuli per session). For discrimination phase the learning criterion was set at 100% correct responses for negative stimuli over 2 consecutive sessions. Each session consisted of 20 positive stimuli and 2, 4 and 6 negative stimuli, increasing from session 1 to 3 respectively. Calves reached the learning criterion after on average 9.67±2.29 (SD) sessions, with no differences between treatments (P=0.16). Training stimuli were then reversed and training continued for 13 sessions. Seven of the 8 calves reared in the social environment were able to reach the learning criterion (after on average 10.29±2.43 sessions), but only one out of seven calves reared individually reached the criterion (after 10 sessions) (Fisher-exact test; P=0.008). This deficit in learning ability may make cattle less able to adjust to changes in management, including the use of new technologies increasingly adopted on working dairy farms.

Developing a titration method to define individual decision difficulty in laying hens

Anna Davies[1], Christine Nicol[1] and Andrew Radford[2]
[1]University of Bristol, Animal Welfare and Behaviour, School of Clinical Veterinary Science, Langford House, Langford, BS40 5DU, Bristol, United Kingdom, [2]University of Bristol, School of Biological Sciences, Woodland Road, BS8 1UG, Bristol, United Kingdom; anna.c.davies@bristol.ac.uk

As a prelude to exploring the relationship between stress and decision-making in hens we aimed to develop a methodology for measuring individual decision difficulty. Decision difficulty can be defined in a variety of ways. Difficult decisions might involve finely-balanced options, have a critical outcome, or require the processing of a large amount of information. For each of these, state-dependant individual factors play a role so decision difficulty cannot be predicted a priori. This study aimed to define the first type of difficulty i.e. finely-balanced options, operationally, using a titration procedure. Fifteen adult laying hens were trained to access a reward, via an unweighted push-door in a T-maze test apparatus. During training, chickens were given a choice between accessing one (Q1) and 6 pieces of sweetcorn (Q6). All 15 hens showed strongly directional preferences for Q6, choosing it at least 90% of the time across 12 consecutive tests and this was defined as an unbalanced decision. During the titration phase weight was gradually applied to the Q6 push door, increasing hens' preference for Q1. The weight at which individuals showed non-exclusive preferences by choosing each option with approximately equal frequency across 12 consecutive tests varied greatly between individuals and was then defined as a balanced decision. We proceeded to examine whether making balanced decisions was more stressful than making unbalanced decisions in a further set of 10 tests per bird. During these tests hens continued to make strongly directional choices when doors were unweighted (Q1-11%, Q6-88%, no choice-1%), and non-exclusive choices when doors were weighted (Q1-36%, Q6-59%, no choice-5%), according to the initial titration. We conclude that this procedure enabled us to define finely-balanced options at an individual hen level, providing a basis to examine whether this potential aspect of decision difficulty is linked with physiological states.

Predicting, preventing, and treating barbering behavior in C57BL/6 mice

Giovana Vieira[1], Amy Lossie[1] and Joseph Garner[2]
[1]Purdue University, Dept of Animal Sciences, 125 S. Russell St., 47907-2042 West Lafayette, IN, USA, [2]Stanford University, Depts of Comparative Medicine and Psychiatry and Behavioral Sciences, 287 Campus Dr., 94305-5410 Stanford, CA, USA; gvieira@purdue.edu

Barbering is an abnormal repetitive behavior characterized by hair and/or whisker plucking. Although it is one of the most prevalent problems in laboratory mice, little is known about its etiology and treatment. The most effective treatment for human compulsive hair-pulling is N-Acetylcysteine (NAC). NAC is the main precursor to glutathione, which protects the brain from reactive oxygen species (ROS). ROS accumulate through cell metabolism, damaging lipids, DNA and proteins within and among nearby cells. We hypothesize that an imbalance in the antioxidant defenses and the production of ROS, through increased cellular metabolism, leads to neuronal damage or cellular quiescence, which manifests as barbering in genetically susceptible animals. We tested this hypothesis by determining the ability of NAC to prevent and cure barbering in n=32 C57BL/6 mice. Animals were assigned into 4 groups: control (regular rodent diet), control diet + NAC (1 g/kg), barbering diet, barbering diet + NAC (1 g/kg). We measured hair-loss biweekly throughout the 24 week study and determined oxidative stress by 4 independent measures in urine, blood and brain. REML logistic regression showed that barbering animals had elevated urinary antioxidant capacity (LR χ^2=18.91; $P<0.0001$) and increased levels of oxidized DNA (LR $\chi^{2=}$4.062; $P=0.0439$), markers of oxidative stress. Compared with controls, NAC protected healthy animals from becoming barbers, and cured animals that were already sick (LR χ^2=5.895; $P=0.0152$), but this effect did not differ between healthy and sick animals (LR χ^2=1.028; $P=0.3106$). Our data suggest a compensatory response from the body to this overwhelming period of oxidative stress, confirm a relationship between oxidative stress and barbering behavior, and provide a potential, highly effective treatment option for hair-pulling behavior (and potentially other abnormal repetitive behaviors) in laboratory mice.

Evaluation of the well-being of rats housed in multilevel caging with red tinted polysulfone

Melissa Swan and Debra Hickman

Indiana University School of Medicine, Laboratory Animal Resources, 975 W. Walnut Street IB 008, Indianapolis, IN 46202, USA; mpswan@iupui.edu

In the laboratory setting, rats are commonly housed in clear caging in brightly lit rooms. As their natural history supports a preference for low-light conditions, this practice is likely stressful for rats. The retinal anatomy of rats suggests that they are unable to see the red spectrum of light, so red-tinted caging would replicate a darkened condition for the rodent. This study examined the welfare of rodents housed in red-tinted caging using a multilevel caging type that allowed the rat to select their microenvironment. Fifty Sprague Dawley rats were divided into five treatment groups: entirely clear, entirely red, red top/clear bottom, clear top/red bottom, and entirely clear with a red intra cage shelter. Rodents were allowed to acclimate to their housing treatment for five weeks before being tested in the elevated plus maze and open field tests. Video recordings were collected for each rat during the peak of the light cycle and the peak of the dark cycle to determine its use of the given environment and the role of color in selection of microenvironment. Results from the video recordings indicated that rats actively sought the red environment when it was provided. Rats within the all red caging showed decreased anxiety within the elevated plus maze ($P<0.04$), while those in a clear cage with a red intra-cage shelter exhibited increased anxiety behavior in the elevated plus maze ($P<0.03$). These findings suggest that an all red cage could be beneficial to the rodent's wellbeing while intra cage shelters can be detrimental.

Comparing daylight behavioural time budgets of young and senescent laboratory mice

Jessica Gimpel and Rommy Von Bernhardi
Centro de Investigaciones Medicas, Escuela de Medicina, Pontificia Universidad Catolica de Chile,
Marcoleta 391, 8330024, Chile; jessica.gimpel@gmail.com

Mice are typically used at young ages in most research experiments, 8 to 12 weeks, unless there is a special reason for using a different age range. Therefore, most guidelines and published biological/ethological data correspond to that period of life. In recent years, modeling human chronic conditions, such as Alzheimer's disease, has led to the development of large colonies of aging mice. Thus, there is an important need to know more about older mice regarding their biology and behaviour to improve care and welfare and to make informed decisions about, for example, humane endpoints, as there is not much specific information about this age. The main purpose of this study was to detect if there was a decline in activity in aged mice. Therefore, we deemed it adequate to observe them during light hours, making it technically easier. We registered behavioural budgets of C57BL/6J mice at three different ages: post-weaning, adult and senescent (1, 3, and over 9 months old respectively). They were housed in conventional cages, 3 or 4 mice per cage (two replicates per age/sex combination, n=41). Behaviour was recorded by direct observation during light hours over 2 weeks. Each group was sampled 10 times, balancing for time of the day. The behaviour of each mouse was registered through instantaneous scan sampling every 10 minutes using a standard published ethogram. Behaviours were subsequently grouped into three broader categories: active, inactive and maintenance. Proportion of time spent in active behaviour was significantly different according to age (H=9.29; df=2; P=0.01). Mann-Whitney tests showed that senescent mice performed significantly less active behaviours (Mdn=0.02; IQR=0.04) than post-weaning (Mdn=0.04; IQR=0.08; U=487; Z=-3.06; P=0.002) and adults (Mdn=0.03; IQR=0.1; U=585; Z=-2.13; P=0.03) respectively. As predicted, active behaviours decrease in senescent laboratory mice in such proportion that differences are detectable even during daylight hours.

Intra- and inter-test consistency of fear tests for adult rabbits

Stephanie Buijs, Luc Maertens and Frank A.M. Tuyttens
Institute for Agricultural and Fisheries Research (ILVO), Animal Sciences Unit, Scheldeweg 68,
9090 Melle, Belgium; stephanie.buijs@ilvo.vlaanderen.be

Validating fear tests by measuring intra-test correlation (i.e. repeatability analysis) is disputed because fear tests are often based on novelty (which decreases with re-testing). However, if test-retest scores correlate positively, re-testing could be used to decrease experimental noise due to disturbances or day-to-day mood fluctuations. Alternatively, testing can be repeated by applying different tests measuring the same underlying construct (i.e. fearfulness), which is evaluated by inter-test correlations. In search of a robust set of indicators of rabbit fear, we studied intra- and inter-test Spearman correlations, using 23 adult females. Per rabbit we performed 4 open-field tests (1 test/day, expected to cause fear due to novelty, social isolation and lack of hiding places), 1 novel-object test (directly after the 4th open-field test) and 4 heart rate recordings (directly after the behavioural tests). The number of open-field squares entered was decreased upon retesting (signed rank test, median day 1: 83 (74-95), day 2: 56 (35-75), $P<0.001$), suggesting that decreased novelty during re-testing affected the rabbits' response. Still, these daily values were highly correlated ($r_s=0.73$, $P<0.001$). The same was observed for post-test heart rate (day 1: 258 (252-267), day 2: 240 (225-249) bpm, $P<0.001$; $r_s=0.50$, $P=0.02$). Although the number of squares entered and post-test heart rate did not correlate on any day ($r_s=0.1$-0.2, $P=0.5$-0.8), both correlated with novel-object contact latency. These correlations were stronger when using a 4 day average than when using day 1 values only (squares: $r_s=-0.61$, $P=0.002$ vs. $r_s=-0.50$, $P<0.015$; heart rate: $r_s=-0.45$, $P=0.03$ vs. $r_s=-0.40$, $P<0.06$), suggesting that re-testing improved the open-field test's reliability. The results suggest that increased open-field locomotion and heart rate are uncorrelated indicators of decreased fearfulness in adult rabbits. However, the observed correlation between locomotion and novel-object contact latency contrasts with previous results for juvenile rabbits, suggesting that such results are age or context specific.

The behaviour during the appetitive motivational state for a reward differ depending on the reward

Claes Anderson[1], Jenny Yngvesson[1], Alain Boissy[2] and Lena Lidfors[1]
[1]Swedish University of Agricultural Sciences, Department of Animal Environment and Health, P.O. Box 234, 532 23 Skara, Sweden, [2]INRA, UMR 1213 Herbivores, ACS, 63122 Saint-Genès Champanelle, France; claes.anderson@slu.se

Learning to identify behaviours indicating positive emotions is important to improve animal welfare. The aim of this study was to identify how an appetitive motivational state (AMS) expresses itself behaviourally. We subjected lambs to a reward (food or opportunity to play) that was preceded by an anticipation phase. We predicted that the lambs during AMS would show more behavioural transitions and increased activity. Forty-two male lambs were allocated into three treatments, resulting in seven pairs of play (P), food (F) and control (C) lambs. Pairs were trained for five consecutive days to enter into a holding pen ($2.7 m^2$) for three minutes before entering a 'reward arena' ($19.5 m^2$) for ten minutes. The reward arena contained two volleyballs and a platform for P and concentrate for F. C only entered the holding pen. After the training period, each pair repeated the procedure twice/week for two weeks during which behaviour was video recorded. Videos were analysed for transitions and durations of behaviours. Data were statistically analysed with mixed generalized model. In the reward arena, P played 31.0% of the time (88.6% of play behaviours was social play), and F spent 85.3% of the time consuming concentrate. In the holding pen, F (46.7 ± 3.5) showed more behavioural transitions than C (37.1 ± 2.2, $P_{FC}<0.05$) whereas P (40.3 ± 3.1) did not differ from C. Both P (12.8 ± 1.0 (percentage of time$\pm$SE)) and F (14.9 ± 0.9) walked more in the holding pen than C (8.1 ± 0.7, $P_{PC}<0.001$, $P_{FC}<0.001$). Exploring did not differ between P (26.6 ± 2.3) and C (29.4 ± 2.0), but F (20.6 ± 2.2) explored less than C ($P_{FC}<0.01$). P (44.6 ± 1.3) were standing still less than C (50.8 ± 1.8, $P_{PC}<0.001$), while F (58.5 ± 4.0) were standing still more than C ($P_{FC}<0.1$). In conclusion, AMS seems to affect the behavioural time budgets prior to receiving a reward and time budgets may differ depending on the type of reward.

Qualitative and quantitative behavioural assessment in sheep during feeding motivation tests

Lindsay Matthews[1], Catherine Stockman[2], Teresa Collins[2], Anne Barnes[2], David Miller[2], Sarah Wickham[2], Else Verbeek[3], Drewe Ferguson[3], Francoise Wemelsfelder[4] and Patricia Fleming[2]
[1]ICT Teramo & Lindsay Matthews Research International, Animal Welfare, viale Mario Torinese 38, Pineto 64025, Italy, [2]Murdoch University, School of Veterinary & Life Sciences, 90 South St, WA 6150, Australia, [3]CSIRO Livestock Industries, Animal welfare, FD McMaster Laboratory, Armidale NSW, 2350, Australia, [4]Scottish Rural University College, Sustainable Livestock Systems Group, Bush Estate, Penicuik, Midlothian EH26 0PH, United Kingdom; lindsay.matthews1@gmail.com

Qualitative Behavioural Assessment (QBA) quantifies expressive behaviour; motivation tests quantify behavioural requirements. We compared qualitative and quantitative measures in feed-motivation tests. Twin-bearing pregnant ewes in three treatments (n=7/treatment; body condition score (BCS) maintained at score 3 (Control), or reduced by 1 BCS unit over 5 wk (Moderate) or 10 wk (Mild) using food restriction) were tested from day 42-67 of pregnancy. Health and welfare was monitored intensively; the sheep in each treatment continued to thrive, did not suffer any long-term effects on health and welfare and produced lambs of similar birthweights. The tests required repeated movement away from and re-approach to a feeder for 4.5 g lucerne chaff. Videos of each ewe during a reward cycle was shown in random order to 11 observers who used their own descriptive terms to score the animals using QBA. There was a high level of consensus between observers (44.7% variation explained, $P<0.001$). A Generalised Procrustes Analysis (GPA) was used to identify the principle dimensions of consensus and variation explained between sheep. Dimension 1 (26.5% variation explained) differed between treatments ($P<0.05$); Control ewes scored low on dimension 1 (calm/bored/comfortable) compared to Mild ewes (more interested/anxious/excited). Dimension 2 (21.4% variation explained) scores were not significantly different between treatments but animals that spent more time 'sniffing and looking for more feed' were attributed lower scores (more hungry/searching/excited, $P<0.05$), than animals that 'did not walk directly to the food reward' (more curious/intimidated/uneasy, $P<0.01$). Dimension 3 scores did not differ between treatments but sheep that had more feeding events during the entire 23-h motivation test were attributed lower scores (more hungry/bold/interested, $P<0.05$) and sheep that ate more had higher scores (more curious/concerned/reserved, $P<0.05$). QBA measures were consistent and identified three main dimensions of behavioural expression. These were in agreement with quantitative measures of feeding, thus supporting the usefulness of QBA as a tool for welfare assessment.

Testing for attention biases in sheep

Rebecca E. Doyle[1], Caroline Lee[2] and Michael Mendl[3]
[1]Charles Sturt University and EH Graham Centre for Agricultural Innovation, Locked Bag 588, Wagga Waggga 2678, Australia, [2]Animal, Food and Health Sciences, CSIRO, Armidale 2350, Australia, [3]School of Veterinary Science, University of Bristol, Langford BS40 5DU, United Kingdom; rdoyle@csu.edu.au

Attention biases are used to identify emotional states in people and animals. An example of this is an anxious individual being easily distracted from a task by threatening cues. The current study was developed to test if attention biases could be identified in sheep. Merino/Border Leister × Dorper wether lambs (n=35) were trained to complete the task of walking to a feed reward over 6 consecutive days. To induce a negative emotional state shearing was performed (n=17) immediately prior to attention bias testing and control sheep did not receive any treatment (n=18). All 35 sheep were tested for attention bias twice by playing known threatening noises while sheep performed the task. The first test used white noise, and the second the noise of a barking dog. Time to complete the task was used to assess distraction. Data were analysed using linear mixed model. Average times for shearing and control groups did not differ (treatment × test interaction P=0.789) with times of 15s and 13s for test one respectively and 19s and 18s for test two (back-transformed means). Despite no evidence of differing attention biases, the noises played caused differences in performance with all sheep completing the task significantly faster on the day of testing (18s±1.2) compared to training results (24s±1.2; t-test: T=4.64, P<0.001). This suggests that the sheep may have found the noises aversive, and so rather than being distracting encouraged the animals to move faster to avoid them. Relative ease of training and the susceptibility of task performance to external stimuli / distractors indicate that the test shows promise and warrants further investigation

Ewes direct more maternal attention toward lambs that express the most severe pain-related responses

Katarzyna Maslowska[1,2], Agnieszka Futro[1,2] and Cathy M Dwyer[2]
[1]Royal (Dick) School of Veterinary Studies, Roslin Institute Building, Easter Bush, Edinburgh, EH25 9RG, United Kingdom, [2]SRUC, ABW, Roslin Institute Building, Easter Bush, Edinburgh, EH25 9RG, United Kingdom; kasia.maslowska@sruc.ac.uk

This experiment was part of a large study on the effects of different castration methods on pain in lambs. It was approved by an Ethics Committee. In this study we investigated whether different methods of castration affected lamb pain behaviours, and the impact this had on the expression of maternal attention by the ewe. Eighty 2-day-old male Mule (Scottish Blackface × Bluefaced Leicester) × Suffolk or Texel lambs were allocated to one of 4 groups (n=20 per treatment): handled only (control, C), castrated using conventional rubber rings (CRR), short-scrotum castration (SSC) and rubber rings castration combined with Burdizzo (RRNC). Duration, frequency and latency of lamb and ewe behaviours were recorded continuously for 30 minutes after the procedure. Lamb behaviours of foot stamping/kicking, tail wagging, head turning, easing quarters, trembling and dog sitting were summed as 'pain-related behaviours' (previously validated as an indicator of pain severity in lambs). Kruskal-Wallis and ANOVA tests were used to determine significant differences between treatment groups, and Spearman's rank correlation test to investigate association between lamb and ewe behaviours. CRR and SSC lambs showed greater active pain-related behaviours (median frequency [Q1-Q3]: C=24.0 [14-42]; CRR=110.5 [63-114], RRNC=25 [13-67], SSC=97 [25-125], H=46.51, $P<0.001$), and restlessness (median frequency of postural changes [Q1-Q3]: C=9.5 [6-17], CRR=33 [28-38], RRNC=10 [7-17], SSC=22 [18-33], H=47.53, $P<0.001$), than C or RRNC lambs. Ewes directed significantly more sniffing and licking towards their lambs that were subjected to CRR and SSC treatments (F3,7=.73, $P<0.001$) and orientated significantly more towards their CRR and SSC lambs, lambs compared to C and RRNC (F3,76=12.88, $P<0.001$). Maternal sniffing and licking behaviours were positively correlated with the frequency of lamb pain- related behaviours ($P<0.001$). Results indicate that ewes are able to distinguish between different levels of pain experienced by their lambs and direct more maternal attention toward those with the most severe reactions.

Stress during pregnancy affect maternal behaviour specifically in passive stress-responsive ewes

Marjorie Coulon[1], Raymond Nowak[2], Frédéric Lévy[2] and Alain Boissy[1]
[1]Inra, UMRH 1213 ACS, Centre de Clermont-Ferrand / Theix, 63122 Saint Genès Champanelle, France, [2]Inra, UMR 6175 Physiologie de la Reproduction et des Comportements, INRA-CNRS-Université de Tours-Haras Nationaux, 37380 Nouzilly, France; marjorie.coulon@clermont.inra.fr

The aim of this study was to investigate the effects of repeated negative experiences in pregnant ewes exposed to various common management practices on the establishment of maternal behaviour. Forty ewes of Romane breed were selected over 120 ewes prior mating according to their behavioural response in a test of isolation/reunion with congeners. The 20 ewes vocalizing and moving the more were defined as active ewes (A) and the 20 ewes vocalizing and moving the less were defined as passive ewes (P). Then during the last third of pregnancy 10 A ewes and 10 P ewes were exposed daily to unpredictable and repeated exposures to uncontrollable and negative challenges, as social isolation, mixing, transport, weighing, delay feeding times (treated ewes) whereas the 20 other ewes were housed without any added challenges (controls). For each ewe its maternal behaviour was observed during the first 30 min post-partum. Then its maternal selectivity was tested 1h30 later by presenting successively an alien and the familiar lamb. At 48 h post-partum its maternal motivation was assessed by using a separation/reunion test in which the ewe could join her lamb despite the presence of a novel object. At parturition, treated ewes tended to emit fewer low-pitched bleats (F=3.01; P=0.06). They made less contact with their lambs (Z=1.92; P=0.05) and tended to vocalized less (Z=1.83; P=0.07) in the maternal motivation test. In addition, behavioural reactivity to stress influenced maternal care: P ewes vocalized less (F=5.61; P=0.02) than A ewes, they spent less time licking their lambs (F=5.02; P=0.03) and treated P ewes expressed more udder refusals (F=4.63; P=0.03). The behaviour of lambs was not affected by prenatal treatment. Therefore, negative emotional experiences during pregnancy affect to some extent maternal behaviour specifically in passive stress-responsive ewes. This study raises questions of potential consequences on behavioural development of the lambs.

Neuro-behavioral reactions to physical stimuli varying in valence in sheep of different mood states

Sabine Vögeli[1,2], Janika Lutz[1,2], Martin Wolf[3], Beat Wechsler[4] and Lorenz Gygax[4]
[1]University of Zurich, Institute of Evolutionary Biology and Environmental Studies, Animal Behaviour, Winterthurerstrasse 190, 8057 Zurich, Switzerland, [2]Research Station Agroscope Reckenholz-Tänikon ART, Centre for Proper Housing of Ruminants and Pigs, Tänikon, 8356 Ettenhausen, Switzerland, [3]University Hospital Zurich, Division of Neonatology, Biomedical Optics Research Laboratory, Frauenklinikstrasse 10, 8091 Zurich, Switzerland, [4]Federal Veterinary Office FVO, Centre for Proper Housing of Ruminants and Pigs, Agroscope Research Station, Tänikon, 8356 Ettenhausen, Switzerland; sabine.voegeli@agroscope.admin.ch

To improve animal welfare, there is need for a better understanding of the interplay of mood and emotion as long- and short-term affective states. We studied this interplay in 29 sheep using neuro-ethological methods. Differential mood was induced by different housing conditions: enriched/predictable conditions for a rather positive, and barren/unpredictable conditions for a rather negative mood. Three physical stimuli were automatically applied to each sheep varying in presumed valence: a negative (pricking without injury), a neutral (slight pressure) and a positive stimulus (massaging). We measured changes in brain hemodynamics using non-invasive functional near-infrared spectroscopy (fNIRS), and locomotor activity using an automatic positioning system as potential indicators of emotional reactions. Linear mixed-effects models were applied and final models were selected in respect to BIC-based model probabilities (mPr). Average deoxy-hemoglobin concentrations [HHb] changed with the course of time throughout the stimulation but independent of stimulus valence or mood (mPr=0.61): [HHb] decreased after the onset of the stimuli and increased after the end of stimulation (maximum average decrease [95% CI]: -0.4 [-0.1;-0.8] μmol/l). No changes were detectable in average oxy-hemoglobin concentrations [O_2Hb] (mPr=0.99). Locomotor activity peaked during stimulation, with the highest values observed during a negative stimulus (mPr=0.89). There was also a weak indication for higher values in sheep from the negative mood group (mPr=0.10; difference from pre-stimulus to stimulus phase: 0.234 [0.161;0.34] to 1.71 [1.15;2.50] m/s; positive mood group: 0.15 [0.10;0.21] to 1.07 [0.73;1.61] m/s). In conclusion, the sheep reacted to all stimuli as revealed by the hemodynamic reaction. They also increased locomotor activity, and most strongly so when given a negative stimulus. Moreover, sheep living under unpredictable housing conditions tended to show stronger reactions to the stimuli. Measurements of locomotor activity and brain hemodynamics were found to be useful and promising indicators of animal affective states useful for studying animal welfare.

Does handling experience alter the response of sheep to the presence of an unfamiliar human?

Susan Richmond, Charlotte Georges, Emma M Baxter, Francoise Wemelsfelder and Cathy M Dwyer
SRUC, King's Building, West Mains Road, Edinburgh EH9 3JG, United Kingdom;
susan.richmond@sruc.ac.uk

Measurement of human-animal relationships is an important component of on-farm welfare assessment. The aim of this study was to validate a voluntary-human approach (VHA) test as a potential method to assess these relationships on sheep farms. Ewes involved in an existing pre-natal stress experiment received one of three handling styles: positive (predictable, gentle voluntary human-animal interactions, n=20), negative (unpredictable, non-contact forced animal movements, n=20) or minimal handling (husbandry only, n=10). Ewes were housed and handled in groups of five. Handling treatments occurred twice daily for ten minutes over five days per week for five weeks during mid-pregnancy. In the VHA test an unfamiliar human entered the ewes' home pens for five minutes twice weekly throughout the course of five weeks. Chi^2 tests were used to analyse quantitative data on the ewes' location and behaviour recorded via scan sampling at 20 second intervals throughout the VHA tests. There were no significant differences between animals at the start of the treatments (week1). Following handling treatments (week 5) significantly fewer positively handled ewes were observed in the back of the pen in comparison to week1 (proportion of ewes in back: week1=80%, week5=40% (χ^2=8.640, d.f.=1, P=0.003). However there were no significant differences in other handling groups between week1 (proportion of ewes in back: 53%) and week 5 (proportion of ewes in back 60% (χ^2=0.075, d.f.=1, P=0.784). During the first 20 seconds of the VHA test in week5 there were significantly fewer individuals involved in aggressive interactions from the positively handled treatment group (5%) compared to those receiving negative or minimal handling (36% χ^2=6.597, d.f.=1, P=0.01). The data suggest positive handling of sheep reduces avoidance of unfamiliar humans and agitated behaviour in their presence. The similarity between negatively and minimally handled ewes may prove problematic for on-farm welfare assessments, especially if animals are handled infrequently.

Attack intensity of pest flies and behavioural responses of pastured dairy cows

Carrie Woolley[1], Simon Lachance[1], Trevor Devries[2] and Renée Bergeron[1]
[1]University of Guelph, Alfred Campus, Animal and Poultry Science, 31 rue St-Paul, Alfred, Ontario, K0B 1A0, Canada, [2]University of Guelph, Kemptville Campus, Animal and Poultry Science, 830 Prescott St., Kemptville, Ontario, K0G 1J0, Canada; cwoolley@uoguelph.ca

The objective of the study was to evaluate the effectiveness of an organic repellent on reducing fly attack intensity and on fly avoidance behaviour, grazing, milk production and stress of dairy cows. Twenty lactating, Holstein dairy cows were randomly divided into 2 groups. Cows were on pasture 24 h/d except at morning and evening milking. During a 9-wk trial period, an essential oil organic fly repellent (2.5% lemongrass, 2.5% geranium, 95% sunflower oil) was manually applied to cows according to a switch-back design, in which groups 1 and 2 switched between treated and untreated each week. Once a week, trained observers performed direct observations for fly counts and direct continuous observations of defensive behaviours (full tail flick, partial tail flick, skin twitch, head throw, leg stamp, lick side, bunching) on each cow. Each cow was fitted once per week with an IGER grazing halter and GPS to electronically record grazing, rumination and travel for a 3-hr period. Milk weights were recorded and milk samples collected to measure cortisol. Cows treated with the fly repellent had significantly lower fly densities in comparison to untreated cows (213±6.8 vs. 69±3.1 flies/cow; $P<0.0001$) throughout the entire trial. Treated cows had lower ($P<0.0001$) rates of tail flicks, skin twitches, head throws, leg stamps and bunching. In comparison to untreated cows, treated cows spent significantly more time grazing (119.1±5.6 vs. 108.1±3.3 min; $P=0.002$), less time ruminating (11.9±2.6 vs. 17.3±2.5 min; $P=0.005$) and travelled longer distances on pasture (0.86±0.05 vs. 0.79±0.03 km; $P<0.001$). There were no differences in milk yield or milk cortisol between treated and untreated cows. The treatment of cattle with an essential oil based fly repellent reduced pest fly attack intensity, cattle defensive behaviours and increased grazing time.

Behaviour and health of different turkey genotypes with outdoor access

Jutta Berk
Institute of Animal Welfare and Animal Husbandry, Friedrich-Loeffler-Institut, Dörnbergstr. 25/27, 29223 Celle, Germany; jutta.berk@fli.bund.de

Intense genetic selection for high growth rate has resulted in heavy turkeys with more health problems and behavioural changes, such as decreased locomotor activity. A veranda and free range area might enhance locomotor activity. This study investigated the effects of three turkey genotypes (B.U.T. 6, Hybrid XL-XL, and Hybrid Grade Maker-GM) and two housing conditions (barn as control, barn plus veranda and free range) on behaviour and health. In two trials, 1,352 one day-old male turkeys were randomly allocated to 12 littered floor pens (each 18 m²). Six groups each were kept without or with veranda (12 m²) and free range (240 m²). Eight pens each contained 54 males (B.U.T. 6, XL) and four pens 61 males (GM). Turkeys were marked individually with transponders and kept for 20 weeks. The use of veranda and outdoor area (OA) as well as health data (mortality, leg posture, walking ability, and pododermatitis) were recorded and analyzed per individual bird using the GLM procedure of SAS*. Significant means were separated using Tukey-Test. Mean time per day in OA were significantly lower for strain B.U.T. 6 (13.4 h/day) compared to XL (16.5 h/day) and GM (15.9 h/day, $P<0.05$). The lightest genotype GM had the least food pad lesions (score 0.93) followed by XL (1.04) and B.U.T. 6 (1.25, $P<0.0001$). Groups with OA had a worse foot pad score compared to control groups ($P<0.0001$). Locomotion and leg posture was significantly influenced by trial and genotype ($P<0.001$). Locomotor activity showed significant genotype differences (GM: 1.47<XL: 1.63<BUT 6: 2.12). Genotype B.U.T. 6 had a worse leg score compared to XL and GM (1.45 vs. 1.27 and 1.32). The results indicate that the lightest genotype GM seems best suited, and the most common genotype in Germany, B.U.T. 6, seems least suited for alternative housing systems.

The body-behavior connection: associations between WQ® measurements and non-cage laying hen behavior

Courtney L Daigle and Janice M Siegford
Michigan State University, Animal Science, East Lansing, MI 48824, USA; lyndcour@msu.edu

The objective was to identify whether Welfare Quality® (WQ) physical measurements can be a proxy measure for hen behavior. Three rooms (135 hens/room) of HyLine® brown laying hens were housed in a non-cage system. Five hens/room (n=15 hens) were marked for individual identification. WQ measurements [foot pad dermatitis (FH), comb pecking (CS), feather damage (FD), and keel bone (KB) score) along with claw length (CL) and body weight (BW)] were taken from 15 marked individuals 2x/month from 18-66 woa. Video observation recorded individual behavior across a 48 hour period at 19, 28, 52, and 66 woa. Linear mixed models identified associations between WQ and behavior. Heavier hens spent more time preening ($P=0.0001$), foraging ($P=0.0001$), resting ($P=0.0067$), feeding at 52woa ($P=0.0281$), and performing 'other' behaviors at 66woa ($P=0.0001$). Hens with longer CL spent more time sitting ($P=0.0001$), less time standing at 28woa ($P=0.0001$) and walking at 19woa ($P=0.0190$) and 66woa ($P=0.0092$). Hens with longer CL spent more time feeding at 52woa ($P=0.0001$) and less at 66woa ($P=0.0252$). Hens with poor CS spent less time dust bathing ($P=0.0022$), more time resting at 66woa ($P=0.0111$), and less time resting at 19woa ($P=0.0032$). Hens with poor FD spent less time standing ($P=0.0001$), resting ($P=0.0286$), and feeding at 52woa ($P=0.0001$). Hens with poor FH spent less time sitting ($P=0.0131$), preening ($P=0.0004$), drinking ($P=0.0353$), and foraging ($P=0.0067$); and more time standing ($P=0.0168$) and feeding at 52woa ($P=0.0001$). Hens with poor KB spent more time resting ($P=0.0136$), less time dust bathing ($P=0.0241$), and less time feeding at 52woa ($P=0.0146$). This analysis identifies behaviors that can be inferred from Good Health and Appropriate Behavior physical measurements collected as part of the WQ. WQ measurements are easier to collect than behavioral observations, so they may identify whether behaviors important to animal welfare are occurring as expected without behavioral observation.

On farm assessment of milking behavior in dairies with automatic milking systems

Janice M. Siegford[1], Gemma Charlton[2], Ali Witaifi[1], Edmond A. Pajor[3], Jeffrey Rushen[2], Jenny Gibbons[4], Elsa Vasseur[5], Anne Marie De Passille[2], Derek Haley[6] and Doris Pellerin[7]
[1]Michigan State University, Animal Science, 474 S. Shaw Lane, East Lansing, MI 48824, USA, [2]Agriculture and Agri-Food Canada, Pacific Agri-Food Research Centre, PO Box 1000, 6947 Highway 7, Agassiz, BC, V0M 1A0, Canada, [3]University of Calgary, Veterinary Medicine, 2500 University Dr. NW, Calgary, AB, T2N 1N4, Canada, [4]Agriculture and Horticulture Development Board, DairyCo, Stoneleigh Park, Kenilworth, Warwickshire, CV8 2TL, England, United Kingdom, [5]University of Guelph, Alfred Organic Dairy Research Center, 10 St-Paul Street, Alfred, ON, K0B 1J0, Canada, [6]University of Guelph, Population Medicine, 50 Stone Rd, Guelph, ON, N1G 2W1, Canada, [7]Laval University, Animal Science, 2425 rue de l'Agriculture, Quebec, QC, G1V0A6, Canada; siegford@msu.edu

An increasing number of dairy farms in North America are using automatic milking systems (AMS) justifying a study on the behaviour of cows under North American farm sizes and management systems. Data were collected from 39 farms across Canada and Michigan. Each farm was visited twice over 5 to 10 d by trained observers. General information from each farm, and data on voluntary visits to the milking robot were collected on 40 focal lactating cows in each herd. Time spent lying down was recorded by accelerometers attached to those same 40 cows. Descriptive data from the survey are presented here. On average (±SD)), farms had 2.18±0.30 AMS with 54.8±1.42 cows per robot. Focal cows visited the robot an average of 2.82±0.15 times per day (min: 1.69±0.11; max: 4.58±0.20). Minimum milking intervals were set at 3.5-8 h, with an average minimum interval of 5.50±0.31 h. The expected milk yield per milking ranged from 6.8-30 kg with an average of 9.78±1.7 kg expected based on number of permitted milkings per day. There were large differences between farms in the average time that cows spent lying down each day (11.2±2.32 h) as well as in the number and duration of lying bouts (10.1±3.95 and 1.2±0.39 h, respectively). Producers had to fetch cows to be milked an average of 3.25±0.34 times per day, on average 11.74±0.32 h had elapsed since the last milking (range=8-16 h). Eleven farms pushed cows out of the AMS if they lingered after milking, and used a mild electric shock twice as often as a blast of air. In general, the data from these North American farms is similar to that reported for cows milked by AMS in Europe; showing large behavioral differences between cows that may reflect their individual adaptation to these voluntary milking systems.

Preliminary evidence of an altered serotonin metabolism in the prefrontal cortex of tail biting pigs

Anna Valros[1], Pälvi Palander[1], Mari Heinonen[1], Emma Brunberg[2], Linda Keeling[2] and Petteri Piepponen[3]
[1]Univeristy of Helsinki, Faculty of Veterinary Medicine, Research Centre for Animal Welfare, Department of Production Animal Medicine, P.O. Box 57, 00014 Helsinki, Finland, [2]Swedish University of Agricultural Science, Department of Animal Environment and Health, Box 7069, 750 07 Uppsala, Sweden, [3]University of Helsinki, Faculty of Pharmacy, Division of Pharmacology and Toxicology, P.O. Box 56, 00014 Helsinki, Finland; anna.valros@helsinki.fi

Serotonin is involved in the regulation of social behaviour, stress reactivity and eating. Tail biting in pigs has been suggested to be related to all these, warranting an investigation of a link to serotonin. The study was conducted on a commercial pig farm with tail biting problems. Pigs were selected based on behaviour as quartets of tail biters (TB, n=13), victims of tail biting (V, n=14), control pigs (non-bitten non-biters) in the tail biting pen (Ctb, n=7) and control pigs in pens with no tail biting (Cno, n=11). The quartets were balanced for age and gender. The pigs were sedated with midazolam, butorphanol and ketamine after which they were blood sampled. They were immediately decapitated on-farm after euthanasia with pentobarbital and their brains were extracted, sectioned and frozen using liquid nitrogen. Prefrontal cortex samples were homogenized and analyzed for 5-HT, and for the main metabolite 5-HIAA. Blood samples were analyzed for amino acids. The ratio of tryptophan to large neutral amino acids (LNAA) and to branched chain amino acids (BCAA) was calculated. Differences between pigs within the categories were tested with ANOVA and correlations between 5-HT, 5-HIAA and 5-HIAA/5-HT and amino acids were tested with Pearson correlations. TB pigs had a higher level of 5-HIAA (4.7 ng/g) (P=0.045) than all other categories (V=3.6; Ctb=3.3, Cno=3.5 ng/g). Amino acid levels did not differ, but TRP, TRP:BCAA and TRP:LNAA tended to correlate positively with the 5-HIAA level, but only in the TB pigs (P=0.06; 0.07 and 0.02 respectively). For TB pigs TRP:BCAA and TRP:LNAA also correlated positively with the level of 5-HT (P=0.004 and 0.008). These results indicate that tail biting behaviour is linked to a changed 5-HT metabolism in the pig brain, and that TB pigs might have an altered tryptophan uptake pattern compared to pigs in other categories.

The effect of grass white clover and grass only swards on dairy cows grazing behaviour and rumen pH

Daniel Enríquez-Hidalgo[1,2], Eva Lewis[2], Trevor Gilliland[3], Michael O'donovan[2], Chris Elliott[1], Dayane L. Teixeira[2] and Deirdre Hennessy[2]
[1]Queen's University Belfast, School of Biological Sciences, 97 Lisburn Road, BT9 7BL Belfast, United Kingdom, [2]TEAGASC, Animal & Grassland Research and Innovation Centre, Moorepark, Fermoy, Co. Cork, Ireland, [3]Agri-Food and Biosciences Institute, Plant Testing Station, Crossnacreevy, BT6 9SH Belfast, United Kingdom; daniel.enriquez@teagasc.ie

The objective of the experiment was to compare grazing behaviour and rumen pH of dairy cows grazing grass only (GR) or grass white clover (GC) swards. Swards were rotationally strip grazed and fresh herbage was offered daily. Sward clover content was measured twice weekly. Eight rumen-fistulated lactating dairy cows (surgery carried out by a qualified veterinary surgeon under license) were arranged into four 2×2 Latin squares and allocated to each treatment for one period each of 14 days: 10 days for acclimatisation and 4 days for data collection. This occurred in May (TS1), July (TS2) and Aug (TS3). Grazing behaviour and rumen pH were measured by fitting the cows with behaviour recorders and indwelling rumen pH probes. The proportion of time spent grazing and ruminating, rumen pH and the proportion of time spent below rumen pH 5.8 were measured. Data were analysed using PROC MIXED in SAS. Clover proportion of GC was 0.08 in TS1, 0.10 in TS2, and 0.28 in TS3. Grazing behaviour was similar between treatments in TS1 (proportion of time grazing 0.41 ± 0.02 and ruminating 0.30 ± 0.01). In TS2 GC cows spent a lower proportion of time grazing (0.42 ± 0.01 vs. 0.46 ± 0.01; $P<0.05$) than GR cows but had similar time ruminating (0.27 ± 0.01). In TS3 GC cows spent a lower proportion of time ruminating (0.28 ± 0.01 vs. 0.33 ± 0.01; $P<0.001$) than GR cows but had similar time grazing (0.40 ± 0.01). There was no effect of treatment on rumen pH (TS1: 5.9 ± 0.08; TS2: 6.0 ± 0.05; TS3: 6.3 ± 0.33), the proportion of time spent below rumen pH 5.5 (TS1: 0.16 ± 0.06; TS2: 0.04 ± 0.02; TS3: 0.04 ± 0.02) or 5.8 (TS1: 0.40 ± 0.08; TS2: 0.22 ± 0.08; TS3: 17 ± 0.06). In conclusion, sward type affected grazing behaviour but only when higher clover proportions were present. Grass white clover swards did not alter the normal rumen pH compared to cows grazing GO swards.

Changes in exploratory feeding behaviour as an early indicator of metritis in dairy cattle

Juliana M. Huzzey, Daniel M. Weary and Marina A.G. Von Keyserlingk
University of British Columbia, Animal Welfare Program, 2357 Main Mall, Vancouver, BC, V6T 1Z4, Canada; jmhuzzey@gmail.com

Animals divide their time foraging between consumption (from known resources) and exploration (for new or better sources); consumption and exploration is predicted to decline during illness. Our objective was to compare feed exploration behaviour and the use of feeding space of metritic and healthy cows. Feeding activity of 101 Holstein dairy cows was measured from 2-wk before until 3-wk after calving using an electronic monitoring system. Pre- and postpartum pens were maintained at 20 cows with free access to 12 feed bins. The number of bins sampled from and amount of feed consumed per bin was used to describe exploration and use of feeding space, respectively. Metritis was diagnosed based on rectal body temperature and vaginal discharge. A sample of metritic cows (n=12) was match paired with 12 cows that remained healthy through to d+21. Data for each cow were summarized weekly for statistical analysis. Differences in feeding activity of healthy and metritic cows were compared using the MIXED procedure of SAS. Metritic cows sampled feed from fewer bins per day than healthy cows during all periods (10.3±0.3 vs. 11.3±0.3 bins/d; $P=0.02$). Similarly, within a meal (feeding bout), metritic cows sampled fewer bins than healthy cows (wk-1: 4.4±0.3 vs. 5.6±0.4 bins/meal; wk+1: 3.3±0.2 vs. 4.0±0.2 bins/meal; $P\leq0.02$). During wk-1, metritic cows consumed less feed than healthy cows from the 3 bins that were most preferred by healthy cows (bins where healthy cows consumed the most amount of feed per day; $P\leq0.05$). During wk+1, there were no preferences for bin location among healthy or metritic cows ($P=0.86$); however, metritic cows consumed less per bin than healthy cows ($P<0.001$). In conclusion, cows with metritis reduced feed exploration behavior and consume less feed from bins preferred by healthier animals.

A search for humane gas alternatives to carbon dioxide for euthanizing piglets: a piglet perspective

Donald C. Lay Jr.[1], Jean-Loup Rault[2], Kimberly A. Mcmunn[1] and Jeremy N. Marchant-Forde[1]
[1]Agricultural Research Service-USDA, Livestock Behavior Research Service, 125 S. Russell Street, West Lafayette IN 47907, USA, [2]Purdue University, West Lafayette, IN 47907, USA; Don.Lay@ars.usda.gov

The identification and validation of a humane method to euthanize piglets is critical to address concern that current methods are not acceptable. This research sought to: (1) identify a method of scientifically determining if pigs find a specific euthanasia method aversive; and (2) develop an innovative method of gas euthanasia. Experiment 1 tested four gas combinations in an approach avoidance test. Pigs were allowed to enter a chamber in which the gases were gradually displacing the air: CO_2, (90%, n=6); $N_2O/CO2$ (60%/30%, n=7); Ar/CO_2 (60%/30%, n=6); and N_2/CO_2 (60%/30%, n=3). Pigs in the CO_2 treatment left the chamber sooner when compared to pigs in Ar/CO_2 ($P<0.001$), N_2O/CO_2 ($P<0.01$) and the N_2 ($P<0.001$) treatments, 3.1±0.2; 9.60±0.4; 8.5±0.6; 9.9±0.1 minutes, respectively. In Experiment, 2 pigs were euthanized using a 2-step procedure using a gradual fill of 1 of 4 gas mixtures (100% CO_2, 70% N_2/30% CO_2, 70% N_2O/30% CO_2 or 70% N_2O/30% O_2. The first step was used to anesthetize, the second to euthanize. When the pig became panicked or anesthetized it was placed into 90% CO_2. The CO_2 pigs had to be moved into the chamber pre-charged with CO2 sooner ($P<0.001$) at 2.9±0.3 minutes, than for the N_2/CO_2 (6.4±0.6 minutes) N_2O/CO_2 (6.7±1.0 minutes), and N_2O/O_2 (14.7±2.1 minutes). All pigs in the N_2O/O_2 treatment were moved into the CO_2 chamber because they became anesthetized. In contrast, none of the pigs in the other treatments became anesthetized and were moved into the CO_2 chamber because they either started to squeal or panic. Although death using nitrous oxide took the longest, pigs did not enter a state of panic, thus this treatment was the most humane.

Zebrafish aversion to chemicals used as euthanasia agents

Devina Wong[1], Marina A.G. Von Keyserlingk[1], Jeffrey G. Richards[2] and Daniel M. Weary[1]
[1]University of British Columbia, Animal Welfare Program, 2357 Main Mall, V6T 1Z4, Canada,
[2]University of British Columbia, Department of Zoology, 6270 University Boulevard, V6T 1Z4,
Canada; devinaww@gmail.com

Zebrafish are increasingly used as a vertebrate model organism for developmental and biomedical research. These fish are commonly euthanized at the end of an experiment with an overdose of tricaine methanesulfonate (TMS), but to date little research has assessed if exposure to this or other agents meets the criteria of a 'good death'. Clove oil and metomidate hydrochloride are alternatives to TMS and have been approved for use in several countries. We used a conditioned place preference paradigm to compare aversion to TMS, metomidate, and clove oil. Zebrafish show a preference for dark or light environments, and by exposing them to anaesthetics in the preferred side of a light-dark box, we were able to show a difference in preference post-exposure. Zebrafish (n=51) were habituated to a testing apparatus where they could choose between a dark and a brightly lit tank. Zebrafish spent approximately 93% of the 900±13.5 s (mean ± S.E.M.) trial in the preferred side. Once habituated, fish were exposed in the preferred tank to biologically equipotent concentrations of the three agents through a 3×3 Latin square design. When retested after exposure zebrafish avoided the preferred tank during the 900-s trial. After exposure to TMS fish spent much less time (591±88 s; mean difference ± S.E.M.) in the previously preferred side. Aversion was less pronounced post- exposure to metomidate (131±68 s) and clove oil (165±97 s) (Kruskal-Wallis test, $P<0.001$). Nine of 17 fish exposed to TMS refused to re-enter to previously preferred side, versus 2 of 18 and 3 of 16 refusals for metomidate and clove oil, respectively (χ^2; $P<0.015$). These results indicate that metomidate and clove oil are humane alternatives to TMS and should be considered when euthanizing zebrafish used for research.

Assessing the humaneness of mechanical methods for killing poultry

Jessica E Hopkins[1,2], Dorothy E F Mckeegan[1], Julian Sparrey[3] and Victoria Sandilands[2]
[1]*University of Glasgow, Animal Health and Comparative Medicine, College of Medical, Veterinary & Life Sciences, University of Glasgow, Bearsden Road, G61 1QH, United Kingdom,* [2]*SRUC, Avian Science Research Centre, Animal Health Group, SRUC, West Mains Road, Edinburgh, EH26 0PH, United Kingdom,* [3]*Livetec Systems Ltd, 1 Sand Road, Flitton, Bedford, MK45 5DT, United Kingdom; jessica.hopkins@sruc.ac.uk*

Poultry are traditionally killed on farm by manual cervical dislocation (MCD). This study aims to design an alternative mechanical method for killing poultry on farm, conforming to new EU legislation (EC1099/2009), which heavily restricts the use of MCD. Three mechanical devices: Modified Armadillo (MARM), Modified Rabbit Zinger (MZIN), a novel mechanical cervical dislocation device (NMCD) and a control (MCD) were tested on anaesthetised birds (induction of 8% Sevoflurane and Oxygen 2 Litres per minute via gas inhalation for approximately 30 seconds) to test efficacy of killing (anatomical damage), and measure time after application for reflexes/behaviours to cease. 230 birds were tested across each kill treatment while encompassing two bird types and ages (broiler/layer × juvenile/adult). GLMM showed kill success was significantly affected by kill treatment ($F=24.50_{(3,222)}$, $P<0.001$), but not bird type, age, or kill weight (means: MZIN=75.0±6.9%; Control=100.0±0.0%; NMCD=96.0±2.3%; MARM=48.7±8.1%). When a kill treatment failed the bird was emergency euthanised and reflex/behaviour data not recorded (33/230 birds failed). The maximum time for all behaviours/reflexes (jaw tone – JT; pupillary – PUP; nictitating membrane – NIC; rhythmic breathing – RB; wing flapping – WF; leg paddling – LP; vent winking – VW) to cease after application was significantly affected by bird type (REML analysis) ($F=39.14_{(1,190)}$, $P<0.001$) (means: broiler=124.5±5.3s; layer=166.2±4.3s), but not device, kill weight or bird age. Kill treatment significantly affected the maximum cease time for JT ($F=18.92_{(3,190)}$, $P<0.001$) (means: MZIN=0.5±0.5s; Control=3.8±1.5s; NMCD=4.6±1.2s; MARM=30.0±7.9s), PUP ($F=51.82_{(3,190)}$, $P<0.001$) (means: MZIN=4.5±2.2s; Control=59.0±3.2s; NMCD=49.8±2.6s; MARM=104.2±12.5s), NIC ($F=2.91_{(3,190)}$, $P=0.036$) (means: MZIN=0.0±0.0s; Control=6.5±2.1s; NMCD=2.3±0.9s; MARM=0.8±0.8s), and RB ($F=2.91_{(3,190)}$, $P=0.036$) (means: MZIN=0.0±0.0s; Control=6.5±2.1s; NMCD=2.3±0.9s; MARM=0.8±0.8s). It was concluded that the most successful device (excluding control) was NMCD in terms of producing sufficient anatomical damage. The reflex/behaviour data suggested that MZIN resulted in birds loosing reflexes/behaviours, which can be indicative of consciousness faster than other treatments. The MARM was shown to be the worst device and will not be taken through to the final experiment, where devices (+control) will be tested on live, conscious birds.

Evaluation of microwave application as a humane stunning technique based on electroencephalography

Jean-Loup Rault[1], Paul Hemsworth[1], Peter Cakebread[1], David Mellor[2] and Craig Johnson[2]
[1]Animal Welfare Science Centre, University of Melbourne, Parkville, 3010 Carlton, Australia,
[2]Animal Welfare Science and Bioethics Centre, Massey University, Private bag 11-222, 4414 Palmerston North, New Zealand; raultj@unimelb.edu.au

Humane slaughter implies that an animal experiences minimal pain and distress before it is killed. Stunning is commonly used to induce insensibility. However, current stunning techniques can lead to variable results or are considered unsatisfactory by some people. Alternative stunning techniques that are quick, humane and acceptable by all are needed. Microwave technology can induce loss of consciousness in rats. High power equipment has now been developed that can focus the energy to produce a rapid rise in temperature in cattle brains (Patent number PCT/AU2011/000527). It is expected that rising brain temperature will stop brain function and result in reversible insensibility. We investigated the effectiveness of different settings for microwave delivery, power and duration, on anesthetized cows to induce insensibility. We used quantitative electroencephalography (EEG) analysis under halothane anaesthesia to assess EEG changes that would be associated with loss of awareness in awake animals. Power and duration were tested based on a stop-go basis to minimize animal use. Due to the ethics implications of that novel technique, 9 cows were used, hence only allowing for descriptive analyses, with each animal acting as its own control. All applications resulted in EEG changes indicative of seizure-like activity, an EEG pattern considered incompatible with awareness. We imposed 5 combinations of 3 different powers and 4 different durations. Shorter duration of application resulted in more rapidly developing EEG changes, with shorter time of onset of EEG suppression (as soon as 3 sec) and shorter time to nadir. Higher power resulted in longer duration of EEG suppression, at least 37 sec and up to 162 sec. Transient bradychardia (average -29%) was observed between 5 and 30 sec post-delivery. Post-mortem autopsies revealed that most histological changes occurred adjacent to the zone of application whereas deeper brain regions showed small to no changes. These preliminary results indicate that the application of microwaves to awake animals would result in the rapid development of insensibility based on the appearance of seizure-like complexes in the EEG. The animal's perception of that particular technique up to the induction of insensibility requires further investigation.

The effect of being outdoors during the day or night on walking and lying behaviour in dairy cows

S. Mark Rutter, Joyce Mufungwe, Norton E. Atkins, Stephanie A. Birch and Liam A. Sinclair
Harper Adams University, Animal Science Research Centre, Newport, Shropshire, TF10 8NB, United Kingdom; smrutter@harper-adams.ac.uk

Previous studies have shown that, when given a free choice between pasture and indoor housing, dairy cows are more motivated to access pasture at night. Giving cows continuous free access between inside and pasture is however difficult to manage in practice. This study investigated the welfare implications of putting high-yielding dairy cows out (i.e. without a choice) on strip-grazed pasture either during the 'day' (mean times 05:11-15:15 h) or the 'night' (15:15-05:11 h) and with or without access to a total mixed ration (TMR) at pasture (without shade in a temperate climate). Forty-eight lactating dairy cows were allocated across four treatments (which were applied at the group level): out during the day with (D+) or without (D-) TMR at pasture, out during the night with (N+) or without (N-) TMR at pasture. All cows had free access to TMR when indoors. Two cows in each of two replicates were fitted with IceTags to record lying, standing and walking over five days. Cows took more steps/hr during the day than the night (149 vs 100; $P<0.001$) and when at pasture compared with indoors (159 vs 91; $P<0.001$), probably due to grazing. There was a significant three-way interaction between measurement period (i.e. day or night), whether on pasture and TMR at pasture ($F_{1,11}=22.20$; $P<0.001$) on the percentage of time spent lying (%L). N- cows showed the lowest %L (during the day), possibly due to them standing eating TMR indoors during the day to compensate for not having TMR access at night. In contrast, D+ showed the highest %L (during the night). The high-yielding cows in groups D- and N- struggled to use grazing to compensate for the lack of TMR at pasture, requiring them to spend longer standing eating TMR when they came indoors, potentially reducing their welfare.

Dairy cow preference for open air exercise during winter under Eastern Canada climatic conditions

Elsa Vasseur[1], François Bécotte[2] and Renée Bergeron[1]
[1]Alfred Organic Dairy Research Center, University of Guelph, Campus d'Alfred, Alfred, K0B 1A0, Ontario, Canada, [2]Institut de technologie agroalimentaire, Campus de La Pocatière, La Pocatière, G0R 1Z0, Québec, Canada; evasseur@alfredc.uoguelph.ca

The Canadian Dairy Code of Practices recommends daily exercise, while organic standards require at least weekly access to open air exercise areas during winter. Management for optimal welfare in winter, when weather conditions may be less than optimal for the animals in Eastern Canada, is not clear. To start building recommendations with regard to outdoor access during winter, we asked the dairy cows what they prefer, by offering them a choice between going out or staying inside, over a range of weather conditions. Thirty-two Holstein lactating cows randomly assigned to 4 groups were submitted to a 6-d preference test repeated 4 times over 8 weeks, consisting in 2 d of forced-choice (3 h) in a free-stall area (home pen, straw-bedded on rubber mat cubicles, slatted flooring) and 2 d in an open air exercise area (pasture field adjacent to the barn), followed by 2 d of free choices (3 h), between the two previous areas. The same feed was available inside and outside. Live observations (feeding from the feeder, lying down, waiting at the fence, time spent inside vs. outside) were conducted every 2 min by scan sampling. Kruskall-Wallis tests were used to analyse at the group level differences on the time spent in each activity in between weeks and hours. Average daily outside temperature ranged from -11 to +7 °C during the experiment and weather varied from very windy with snow to sunny. Snow covered the ground during week 1-2 but disappeared for week 3-4. All cows chose to be outdoors for at least 1 h and 50 (week 1) to 100% (week 3-4) chose to stay outdoors for 3 h, even with snow coverage and negative temperatures. Cows spent 45-70% of their first hour outdoors at the feeder, suggesting that feed should be provided while cows are using an open air exercise area. The experiment is currently repeated with Canadian and Jersey cows housed in deep-bedded composted pack.

Dairy heifer preference for being indoors or at pasture is affected by previous experience

Priya R. Motupalli, S. Mark Rutter, Emma C. Bleach and Liam A. Sinclair
Harper Adams University, Animal Science Research Centre, Newport, Shropshire, TF10 8NB, United Kingdom; pmotupalli@harper-adams.ac.uk

Several factors influence dairy cow preference for pasture. Increasing interest in continuously housed systems makes it important to understand these factors to ensure dairy cow welfare is not compromised. This study aimed to determine if previous experience of pasture influenced preference for pasture vs. housing. Twenty four heifers (n=8 for statistics) were reared in one of two treatment groups: exposure to pasture May-November 2011 (P) vs. zero exposure to pasture (Z). Preference for pasture was tested at 16.4±1.2 months (mean ± SD) during July-September 2012. P and Z heifers were tested in 2 groups of 3 for 10 days, with 4 repetitions. Heifers had access to one of two adjacent areas of housing (85.1 m^2, 1.3 free stalls/heifer). *Ad libitum* haylage and 2 kg concentrate/ heifer (once daily) were available indoors. Heifers had continuous access to one of two adjacent 0.22 ha pastures via a 38 m track. Location, posture, jaw activity, and investigation of grass were recorded using scan sampling on days 1, 3, 6 and 10 of each measurement period during 06:00-22:00h. A one-way ANOVA revealed that Z heifers spent more time indoors (79.0 vs. 54.9%±4.71, mean ± s.e.m., $P=0.001$), less time at pasture (13.6 vs. 37.0%±4.82, $P=0.002$), more time eating hay (24.2 vs. 17.0%±1.36, $P<0.001$), less time grazing (3.5 vs. 18.1%±2.04, $P<0.001$), more time idling (46.1 vs. 38.4% ±1.33, $P<0.001$) and more time investigating grass (5.07 vs. 0.160%±0.64, $P<0.001$) than P heifers. Z heifers preferred to be indoors (absolute preference and cf. P heifers). Z heifers spent less time grazing and more time investigating grass which suggests there is a possible learned component to grazing.

Motivation for access to pasture in dairy cows

Andressa A. Cestari[1,2]*, Jose A. Fregonesi*[1]*, Daniel M. Weary*[2] *and Marina A. G. Von Keyserlingk*[2]
[1]*Universidade Estadual de Londrina, C.A.R.E. – Cuidado Animal e Responsabilidade Ética, Rod. Celso Garcia Cid PR 445 Km 380, Campus Universitário, 86051-980, Londrina, PR, Brazil,* [2]*Faculty of Land and Food Systems, University of British Columbia, Animal Welfare Program, 2357 Main Mall, Vancouver, V6T 1Z4, Canada; awandressa@yahoo.com.br*

Access to pasture is thought to be important for dairy cows' welfare, but no work to date has directly assessed the strength of motivation for indoor-housed cows to access pasture. This study used a weighted push-gate to measure the strength of motivation for pasture access. Twenty-four Holstein dairy cows, averaging ± SD 220.5±14.4 DIM and 30.0±3.8 l of milk/day, were randomly assigned to four groups (6 cows each). Each of these groups was tested separately in each of two experiments. In Experiment 1 of the study cows were allowed to push a weighted (7 kg) gate after the morning and afternoon milking to access fresh TMR. Once all cows were trained to perform this task the weight was increased daily (in 7 kg increments) until they were no longer willing to perform the task. In Experiment 2 cows were trained to access pasture using the same weighted gate and motivation was assessed in a free-choice test. Again the weight was increased daily until cows refused to perform the task. Differences in maximum weight pushed were compared within cow using a paired t-test. Cows pushed up to 36.54±14.09 kg to access the TMR versus 31.63±17.54 kg to access pasture (t_{22}=-1.34; P=0.1958). During daylight hours cows spent 76.01±30.55% of their time outdoors (with the remainder spent inside the freestall barn). While on pasture during the day cows spent 40.66±19.42% of the time grazing. These results indicate that dairy cows normally housed indoors are as motivated to access pasture as they are to access fresh TMR after milking.

Observational study of the presence and characteristics of bruises in cattle carcasses at the slaugh

Stella Huertas[1], Gonzalo Crosi[1], Juan Imelio[2], José Piaggio[1,3] and Andres Gil[1,3]
[1]Facultad de Veterinaria, Universidad de la Republica, Instituto de Biociencias, Lasplaces 1550, 11600 Montevideo, Uruguay, [2]National Meat Institute, Ricon 545, 11000 Montevideo, Uruguay, [3]Ministry of Livestock Agriculture and Fisheries, Epidemiology Unit, Cosntiutyente 1476, 11100 Montevideo, Uruguay; gonzalocrosim@gmail.com

Welfare is becoming an increasingly important issue world-wide. Understanding the natural behaviour of animals is important in production systems. This helps avoid injuries and/or illnesses caused by stress and incorrect handling, as well as to understand the effects of human-animal interactions. Improving the way animals are treated will improve their welfare and production. Beef cattle are both herd and prey animals. On arrival at slaughter plants they must be handled respecting their natural behavior. This implies a smooth handling, without shouts or any physical contact, through adequate facilities according to the number and size of the animals. Otherwise, there will be a detriment in the welfare of the animals, carcass bruises and decrease of meat quality. The objective of this study was to characterize the welfare of beef cattle through some indicators such as post-mortem lesions found in the carcasses. Seven slaughter plants authorized for export were visited in Uruguay during October and November 2012. An observational study of the carcasses was performed, corresponding to 25% of the slaughter of each visit day. Carcass bruises were recorded by trained observers taking into account its location, depth and shape, according to a previously standardized protocol. The results showed that from a total of 1030 carcasses observed, 44.4% of them had bruises, half of them had one lesion and the rest had two or more lesions; thus, one of two carcasses showed bruises. The characterization of the prevalence of these bruises is a valuable tool to be used routinely in slaughter plants, providing information on the level of welfare of animals and the way they are treated during the pre-slaughter period.

Which measures of acceleration best estimate the duration of locomotor play by dairy calves?

John Luu[1], Julie Føske Johnson[2], Anne Marie De Passillé[3] and Jeffrey Rushen[3]
[1]University of British Columbia, Vancouver, BC, Canada, [2]National Veterinary Institute, Oslo, 0106, Norway, [3]Agriculture and Agri-Food Canada, Agassiz, BC, Canada; Jeffrushen@gmail.com

Measures of acceleration have been used as automated measures of activity. We examined which measures of acceleration were best correlated with locomotor play and how the sampling rate of the accelerometer affected the correlations. Tri-axial accelerometers were attached to 28 6-7 week-old Holstein calves, who ran freely 10 times in a large arena for 10 mins. We correlated measures of acceleration (summed acceleration, and the number of peaks in acceleration, when acceleration was measured 33/s, 11/s and 1/s) with the frequency and duration of running, jumping and walking (scored from video-recordings). Peaks in acceleration occurred most often when the calves were running. When all running trials (n=271) were used, the overall number of peaks in acceleration measured at 33/s was correlated highly with duration of running (r=0.96), the frequency of jumping (r=0.96) and the ratio of locomotor play (running and jumping) to walking (r=0.75). However, a sizeable correlation (r=0.90) was also obtained between the duration of running and the frequency of peaks in the forward direction at a 1/s sampling rate, which would allow measures of acceleration to be taken for 18h. During a single running test, the number of peaks in the forward direction sampled at 1/s was significantly correlated across calves with the duration of running (r=0.90 n=28 $P<0.001$), the frequency of jumping (r=0.78 n=28 $P<0.001$) and the milk intake of the calves during that day (r=0.53 n=28 P=0.004). Accelerometers provide good estimates of the locomotor play of calves and can distinguish this from walking. Reducing the rate at which acceleration is measured lengthens the time that measures can be taken without greatly reducing the accuracy of the estimates.

Determination of suckling process patterns affecting the growth of piglets using data mining methods

Janko Skok and Dejan Škorjanc
Faculty of Agriculture and Life Sciences, University of Maribor, Department of Animal Science, Pivola 10, 2311 Hoče, Slovenia; janko.skok@um.si

The complexity of lactation period is determined by specific behaviour (social hierarchy, intense competition for suckling position) and physiological demands of neonatal piglets (suckling-milk, thermoregulation). The aim of the present study was to analyse and elucidate piglets' suckling behaviour and suckling process using data mining methods to determine patterns and factors that affect piglet growth performance. Eighteen sows with their litters (180 piglets) were included in the analysis. Throughout the lactation period there were 27 observations per litter – two to four consecutive sucklings on 1, 4, 7, 10, 14, 18, 21 and 25 d, and on the last (28) day of lactation (i.e. equivalent to the age of piglets). Interactions between behavioural parameters, maternal and litter properties were analysed with data mining methods (WEKA 3.6.8) using classification and regression decision trees. Preliminary results showed strong importance of three interacted behavioural components on the average daily gain (AWG) and 28 day weaning weight (BW_{28}). The most important factors were the number of unique teats exploited by a piglet and the position of piglet on the sows' udder. Suckling position at the anterior udder and lower number of unique teats exploited by a piglet led to the higher AWG and BW_{28}. Suckling stability, which appeared at the second level of the model, was also significantly involved in determination of the growth performance ($P<0.01$). Proportion of multi-teats suckling's also appeared in the models, but often on the third or lower level and thus does not have direct impact on the growth performance. Results indicated, that all of the crucial behavioural components were strongly influenced by the parity of sow, i.e. unique teats exploited ($r_{reg.model}=0.77$) as well as suckling stability (73% correctly classified instances in the classification tree) and multi-teats suckling's (87% of correctly classified instances). Models have emphasized parity as the most deciding attribute and showed strong influence of mother on the suckling behaviour of piglets. Piglets from the first-parity gilts showed significantly lower stability of suckling ($P<0.0$ 1), highest number of exploited teats ($P<0.01$) and consequently significantly lower growth performance ($P<0.05$).

Estimates of genetic parameters for flight speed in Île-de-France and Île-de-France crossbred sheep

Priscilla Regina Tamioso, Laila Talarico Dias and Rodrigo de Almeida Teixeira
Federal University of Parana/Universidade Federal do Paraná – UFPR, Departamento de Zootecnia, Rua dos Funcionários, 1540, 83035-050, Curitiba, Paraná, Brazil; priscillatamioso@gmail.com

Flight speed (FS) is one of the most used behavioral parameters to assess reactivity in livestock species, as it is considered an objective and easily measured indicator, although relatively few publications have been reported for sheep. Thus, the aim of this study was to estimate heritability and repeatability for flight speed in 106 Île-de-France and Île-de-France × Texel crossbred lambs, offspring of 67 dams and 2 sires, assessed at 30, 60 and 90 days old, on average, during 2012. The (co)variance components were obtained using REML procedure in univariate analysis. The model included the fixed effect of contemporary group, formed by the month of birth, sex and type of management (except at 30 and 60 days old, since all the animals belonged to the same group) and, as a covariate, the linear effect of the age of dam. In order to estimate the genetic parameters, only the direct genetic effect was considered in the model. Heritability coefficients for FS at 30, 60 and 90 days of age were 0.64±0.50, 0.39±0.37 and 0.00±0.00, respectively, indicating low possibility of direct selection, due to the high standard errors. Likewise, the repeatability estimate was equal to 0.09±0.07, low in magnitude, probably as a function of the small number of animals. The results show that, at 30 days of age, it may be possible to identify variability among the lambs assessed through their behavioral responses in flight speed. Nevertheless, to obtain genetic progress for FS, it would be necessary to perform direct selection for this trait, although the low coefficients suggest a slow progress.

Calves can be taught to urinate in a specific place

Alison Vaughan[1,2], Anne Marie De Passillé[1], Joseph Stookey[2] and Jeffrey Rushen[1]
[1]Agriculture and Agri-Food Canada, Ethology, 6947 Highway 7, V0M 1A0, Canada, [2]University of Saskatchewan, Saskatoon, Saskatchewan, S7N 5B4, Canada; alison.vaughan@usask.ca

The accumulation of faeces and urine in dairy barns is a cause of cattle and human health issues and environmental problems. It is usually assumed that cattle are not capable of controlling urination and defecation. We tested whether cattle could be taught to urinate in a specific location. Six female Holstein calves, each with a yoked control, (31-68 d of age), were habituated to an experimental stall. On d1 experimental calves were held in the stall and injected IV with a diuretic (Salix at 0.5 ml/kg BW) and immediately upon urinating were released from stall into a sawdust bedded reward pen where they received 250 ml milk. On the subsequent 16 d calves were returned to the stall. If they urinated without the diuretic, they e received the milk reward and on the next day were re-tested without a diuretic. Calves that did not urinate within 15 min exited the stall into a small unbedded 'time out' pen, were held there for 5 min and received the diuretic the following day. Yoked control calves were never given diuretic but held in the stall for the same amount of time and received the same 'reward' or 'punishment' as their matched treatment animal had been the previous day. We used the frequency of 'voluntary' urinations and defecations that occurred in the stall on days without a diuretic to compare treatment calves and their yoked controls. Experimental calves had a higher frequency of 'voluntary urinations' (i.e. those not induced by diuretic) than their yoked controls (mean±SE = 5.25±0.95 vs 2.32±0.52; matched pairs t test $P=0.02$). There was no difference between treatments in the frequency of defecation. Our results show that it may feasible to train cattle to urinate in specific areas of the barn. Further research is needed to determine whether the higher frequency of 'voluntary urinations' by calves in the training treatment reflects operant conditioning or whether the calves learn to associate the stall with the diuretic through classical conditioning.

Predictability reduces stress on fish

Patrícia Tatemoto[1], Caroline Marques Maia[2], Graziela Valença Da Silva[2], Mônica Serra[2] and Gilson Luiz Volpato[1,2]
[1]CAUNESP, UNESP – campus de Jaboticabal (SP), via de Acesso Prof. Paulo Donato Castellane, s/n, 14884-900, Brazil, [2]IBB, UNESP – campus de Botucatu (SP), Fisiologia, distrito de Rubião Júnior, s/n, 18618-970, Brazil; carolmm_luzi@hotmail.com

Animals prefer conditions where they can predict time they will be subjected to a stressor. This was shown to reduce intensity of the response to stress in mammals. To know whether the effect of predictability on stress might be a general phenomenon in vertebrates, here we tested the effect of a predictable stressor (confinement) on ventilator frequency of the territorial fish Nile tilapia (*Oreochromis niloticus*). Adult fish, irrespective of sex, were subjected to different schedules of predictability of a confinement stressor (restricted to 10% of the aquarium for 30 min, using an acrylic opaque plaque) and the ventilatory frequency was measured during four consecutive days and the morning of the 5th day. Fish were individually confined according to four treatments: daily at fixed times (08:00 h and 13:00 h; predictable treatment); at random, once in the morning and once in the afternoon (unpredictable treatment); only in the morning (08:00 h) of the 5th day (totally unpredictable treatment); and not confined fish (control treatment). Ventilatory frequency was measured in the morning of the 1st and the 5th day: 5 min before the confinement (pre-stressor) and 0, 15, 35 and 75 min after removing the stressor (post-stressor). We found that confinement increased ventilatory frequency in all treatments, but this response was higher in fish subjected to a totally unpredictable confinement (ANOVA for repeated measures, $P<0.05$). As ventilatory frequency has been used as an indicator of stress, our findings indicate that predictability reduces stress in fish, thus enlarging the participation of predictability of stressors in vertebrates. In practical terms, when a stressor is necessary to be applied on the fish, it should be predictable, so that its impact might be reduced. Funding agency: CNPq.

Selection for wheat digestibility affects emotionality in broiler chicks

Aline Bertin[1], Alice Pelhaitre[2] and Sandrine Mignon-Grasteau[2]
[1]INRA, CNRS UMR 7247; INRA UMR85 Physiologie de la Reproduction et des Comportements, 37380 Nouzilly, France, [2]INRA, Unité de Recherche (UR) 83 Recherches Avicoles, 37380 Nouzilly, France; aline.bertin@tours.inra.fr

Genetic selection for high productivity in poultry may be associated with risks to animal welfare. In this study, we tested whether divergent selection for a new trait linked to feed efficiency, i.e. digestibility, had correlated side effects on the behaviour of broiler chickens. Digestibility was assessed from apparent metabolisable energy values corrected to zero nitrogen balance (AMEn). We tested the behaviours of 80 broiler chicks from a divergent selection experiment (tenth generation) for high (D+ line, 40 chicks) or low (D- line, 40 chicks) wheat digestibility. Birds were tested between 6 and 10 days of age. We used tonic immobility (TI) and open-field tests to evaluate the emotional reactivity of the birds. Fear of a novel food was also assessed. D+ birds had significantly longer duration of TI and latency of the first step in the open-field tests than D- birds (TI duration D+ vs. D- birds: 78.2±14.5 s vs. 36.6±8.4 s, Mann-Whitney, $P<0.01$; latency of first step: 63.6±9.8 s vs. 40.2±7.1 s., Mann-Whitney, $P<0.05$). Our data thus suggest a higher emotional reactivity in D+ birds than in D- birds. On the other hand, D+ birds showed significantly lower latencies to approach a novel food than D- (78.5±19.8 s vs. 110.84±21.4 s, Mann-Whitney, $P<0.05$). Weight gain from hatching to 24 days did not differ significantly between the two lines. Selection indirectly affected different dimensions of fear behaviour. D+ birds expressed higher inhibition in novel environments than D- birds. However, in the presence of a novel food the opposite behaviour was observed. Behavioural differences as a result of divergent digestibility could potentially be due to 'co-selection' or genetic drift. Further investigations are needed to understand the underlying mechanisms and be able to select animals with genotypes associated with traits beneficial for both the environment and animal welfare.

The science of animal sentience – be part of the movement

Gemma Carder, Helen Proctor and Rosangela Ribeiro
WSPA, World society for the protection of animals, 222 grays inn road, WC1X 8HB, UK, United
Kingdom; gemmacarder@wspa-international.org

Animal sentience is a subject of growing international research interest. Understanding of animal sentience underpins the entire animal welfare movement; understanding of sentience in a range of species is crucial in consideration of how we as human beings should treat animals. The World Society for the Protection of Animals (WSPA) is committed to promoting animal sentience science as a mainstream and credible science in the field of animal welfare. As a result in August last year, WSPA launched an interactive website called the Sentience Mosaic (www.sentiencemosaic.org) which acts as a resource for students, scientists, academics and anyone whose work may involve animals. The site promotes and shares scientific research on animal sentience and features interviews with leading scientists. The Sentience Mosaic also offers the opportunity for anyone to become actively involved through monthly online debates and a forum. The Sentience Mosaic is currently available in English and Spanish and will be launched in Portuguese early this year. WSPA aims to promote animal sentience science through the Sentience Mosaic and other avenues. In this way WSPA aims to reach out not just to individuals and organisations working in animal welfare, but also to others whose work may affect animals. This may include encouraging those involved in conservation, agriculture, and animal experimentation to consider animal sentience in their work. Looking forwards we are optimistic that the future of animal sentience research is promising. There is still a great deal to learn about the capabilities of non-human animals. We hope that future research explores sentience not just in vertebrates, but also in invertebrates, in an ethical and humane way.

Effects of rearing practices on the behaviour of dairy calves

Lívia Carolina Magalhães Silva, Maria Fernanda Martin Guimarães, Luciana Pontes Da Silva and Mateus J. R. Paranhos Da Costa
Faculdade de Ciências Agrárias e Veterinárias, Unesp, Campus Jaboticabal-SP, Animal Sciences, Via de acesso Professor Paulo Donato Castellane, s/n, 14884-900, Brazil; lmagalhaesilva@gmail.com

The aim of this study was to assess the effects of rearing methods on the behaviour of dairy calves. Male calves (48 Holstein-Jersey and Hostein-Gir crossbred) were followed from birth to 120 days of age. Calves were allocated to one of two treatments: (1) conventional, in which the calves were kept in individual shelters until weaning (which occurred abruptly), tied with a 2 m chain, fed milk direct from a bucket and had few interactions with the handler; and (2) Alternative, which incorporated a set of rearing practices, including group housing, suckling in buckets with nipples, frequent social interaction with the handler (including tactile stimulation) and gradual weaning. Calves were tested for flight distance test and docility as assessed by the latency for the calf to attempt escape, time required to restrain a calf in a corral corner and the time the calf accepted contact by the handler. Analyses were with the MIXED procedure in SAS ($P<0.05$). There were no significant differences for latency for the calf to attempt escape between treatments (6.67 ± 3.15 vs 5.84 ± 2.65 s, $P=0.7140$). However, calves reared in the alternative rearing treatment had shorter flight distances (1.00 ± 0.09 vs 0.48 ± 0.09 m, $P=0.0001$), required less time to restrain (75 ± 13.43 vs 45 ± 11.35 s, $P=0.006$) and were more willing to tolerate contact by the handler (37.00 ± 6.03 vs 54.00 ± 4.97 s, $P=0.0013$). These results illustrate that modified rearing practices can increase docility and handling ease in dairy calves. Financial support: FAPESP

Tear staining as a potential welfare indicator in pigs

Shelly P. Deboer and Jeremy N. Marchant-Forde
USDA-ARS, Livestock Behavior Research Unit, Purdue University, 125 S. Russell St., IN 47907, USA; marchant@purdue.edu

Facial tear-staining in rodents increases with environmental stress and in pigs, increases during social isolation. We aimed to determine if tear-staining is related to other classical indicators of stress, such as behavior, heart rate (HR) and plasma cortisol concentrations. Sixteen castrated male weaner pigs were each studied over a 2-week period; on day 1, each pig was housed individually with access to 4 adjacent enriched pens (1.5×3.0 m), from 1 of which it could see a companion pig in an un-enriched pen across the passageway. On day 8, the pig moved across the passageway to become the companion pig. They had *ad libitum* access to food and water. Behavior of each pig was recorded continuously. On days 1 and 8, pigs' faces were wiped clean. Photos were taken on days 1-3, 6-10, 13-14 to assess tear staining by direct measurement of stain area. HR was recorded for 8 hours on days 2, 6, 9 & 13 and blood samples were taken by jugular venipuncture on days 3, 7, 10 & 14 for cortisol analysis, so as not to influence HR data. Data were analyzed using Wilcoxon signed-rank tests and Spearman's rank correlation. Tear-staining score and area increased between d1 and d7 and again between d8 and d14 ($P<0.001$), but was not different between enriched pen and companion pen periods. Plasma cortisol and measures of HR and HR variability (HRV) did not differ over time. Tear-staining area was negatively correlated with plasma cortisol concentrations on d3, d7 and d10 and with time spent lying laterally on d7 (all $P<0.05$), but positively correlated with the sympathetic nervous system indicator (SNSI = low frequency power/high frequency power) on d3 and d6 (both $P<0.05$). The results indicate that tear-staining may provide information about the relative activation of the HPA and SAM axes, and hence with further study, may provide information about individual pigs' welfare.

Environment and coping with weaning affect immune parameters of piglets 25 days after weaning

Marije Oostindjer[1], Marjolein Priester[2], Henry Van Den Brand[2], Henk Parmentier[2], Ger De Vries Reilingh[2], Mike Nieuwland[2], Aart Lammers[2], Bas Kemp[2] and J. Elizabeth Bolhuis[2]
[1]Norwegian University of Life Sciences, Department of Chemistry, Biotechnology and Food Science, P.O. Box 5003, 1432 Aas, Norway, [2]Wageningen University, Adaptation Physiology Group, Centre for Animal Welfare and Adaptation, P.O. Box 338, 6700 AH, the Netherlands; marije.oostindjer@umb.no

Housing can have long-lasting effects on behaviour, but the long-term effects of past environment and stressful events such as weaning and a change in environment on immune parameters in pigs are unknown. To study this, piglets from barren (B, 8 litters) and enriched (E, 8 litters) preweaning pens were weaned at 30.2 ± 2 days, mixed, and housed in B or E postweaning pens in a 2×2 design (4 treatments, n=8 pens per treatment, 4 pigs per pen, two rooms, two batches). Blood samples (5ml) from the jugular vein were collected 24 days postweaning. We determined natural antibody (KLH-NAb) titres, classical and alternative complement (second batch), and neutrophil/lymphocyte (N/L) ratio (second batch). As indicators of coping with weaning, postweaning growth and diarrhea were measured over 14 days. Food intake and play behaviour were scored live on 4 days postweaning (720 scans per piglet).Immune parameters were analysed using generalized linear mixed models, including preweaning, postweaning environment, their interaction, batch and room as main factors, and pen nested within pre- and postweaning environment and room as random effect. Spearman rank correlations were performed on the residuals of a GLM including pre- and postweaning environment and batch as main effects. Preweaning barren pens showed higher complement than preweaning enriched pens (classical: $P=0.09$; alternative: $P=0.04$). Piglets that changed from enriched to barren environment at weaning had the highest N/L ratios ($P=0.04$). In addition, a reduced ability to cope with weaning correlated with higher IgM titres (postweaning growth, r=-0.18, $P=0.05$, play behaviour, r=-0.18, $P=0.04$), higher classical complement activity (diarrhea, r=0.35, $P=0.007$) and N/L ratio (time spent eating on day 1 postweaning, r=-0.23, $P=0.08$). The past environment and the effects of stressful event such as switching environment and weaning are still visible in immune parameters 24 days postweaning. Future research should focus on the long-term effects on welfare.

Recumbency and loss of the pedal withdrawal reflex as indicators of insensibility in mice

Carly Moody, I. Joanna Makowska, Beverly Chua and Daniel Weary
University of British Columbia, Animal Welfare Program, 2357 Main Mall, V6T 1Z4, Vancouver BC, Canada; carly_moody@hotmail.com

Rodents are commonly euthanized using an overdose of isoflurane followed by CO_2, or by CO_2 alone. Both involve anesthetizing to unconsciousness, then switching to a high flow rate of CO2 to complete the euthanasia procedure. It is important that rodents are unable to experience pain when the switch is made, as exposure to high concentrations of CO_2 causes pain. Animals may still be able to experience pain even after recumbency, but there is little consensus on behavioural indicators of insensibility in anesthetized animals. We measured recumbency, loss of the righting reflex and abolishment of the pedal withdrawal reflex to assess depth of anesthesia for mice (n=7) euthanized using isoflurane followed by a high flow rate of CO_2, and mice (n=6) euthanized using CO_2 alone. Once mice were recumbent they were tested for loss of the righting reflex, then toe pinches were given on alternating hind limbs every 10 s until three consecutive non-responses occurred. Six of the 7 mice tested using isoflurane moved when tested for the righting reflex and all showed purposeful movement after loss of the righting reflex occurred; these responses were never observed when using CO_2 alone (Binomial test; $P<0.01$ for both measures). All mice tested with isoflurane showed a pedal-withdrawal response versus only one mouse tested with CO_2 alone ($P<0.005$). These results indicate that recumbency cannot be used to infer insensibility for mice anesthetized with isoflurane. The interval between onset of recumbency and loss of the pedal reflex for mice exposed to isoflurane averaged ± S.D. 40.4±12.9 s. Thus we recommend animal users wait at least 90 s after the appearance of recumbency, before exposing mice to a high flow rate of CO_2.

Location of shading structure on paddock and its effects on use of shade by dairy cows

Steffan Edward Octávio Oliveira, Alex Sandro Campos Maia, Cíntia Carol Melo Costa and Marcos Chiquitelli Neto
São Paulo State University, Animal Science, Professor Paulo Donato Castellani Access route w/n, 14883-900, Brazil; steffan_edward@yahoo.com.br

The aim of this study was to determine whether the use of shade is affected by their distance from the trough and drinking fountain, and if there is a relation between time of use of shade with its ability to block radiation. Ten Holstein cows divided into five groups of two animals were used, being one with predominantly white coat and other black, distributed into five treatments. T0=no shade; T30=artificial shade with 30% of solar radiation blockage; T50=50%; T70=70% and T100=100%. The group was the experimental unit on this study. Each treatment was allocated in different paddocks and all groups were evaluated in all treatments. The work was divided into two phases. On the phase 1 the distance from shade to the trough and drinking fountain was approximately 40 meters, while on phase 2 this distance was reduced to five meters. In both phases each group stayed in each treatment for five days. Meteorological and behavioral data were recorded from 8:00 to 17:00 and the production of two daily milking. Data were subjected to analysis of variance using the least squares method, and the adjusted means were compared by Tukey test ($P<0.05$). The results showed that on phase 1 the cows did not use shade, ($0,14\pm1.23\%$), preferring to stay close to the trough and drinking fountain under solar radiation than walk to the shade, while on phase 2 they spent on average 26% ($26.34\pm1.17\%$) of the time on shade. This may explain why we found a significant difference ($P<0.05$) in milk production between phases (1=10.29 l/milking; 2=10.76 l/milking). However there was no significant difference for time on shade between treatments (T30=11.57 ± 1.87; T50=12.98 ± 1.87; T70=20.01 ± 1.87; T100=23.23 ± 1.87). We also found a significant difference for use of shade according to the coat color. Animals with white coat used shade for approximately $14,36\pm1,18\%$, while those with black coat used for $12,13\pm1,18\%$ of the time. The location where we provide shade to cows can affect the time of use of this resource and production.

A mathematical modelling approach to study stress behavior patterns on brown capuchin

Adriano Gomes Garcia, Josemeri Aparecida Jamielniak and Magda Silva Peixoto
São Paulo State University, Biostatistics Department, Rubião Jr District, Botucatu, SP, 18618-970,
Brazil; adrianogomesgarcia@gmail.com

Exhibition of species-typical behavior patterns has long been invoked as a key criterion in evaluating the welfare of captive animals in zoo and research settings. Some factors that can influence on stress are loud sounds and food availability. In this work, we have developed a mathematical model that incorporates Fuzzy Set Theory to study stress behavior patterns on brown capuchin; whose input variable are 'Visitor numbers', 'Frequency of periodic loud sounds at the zoo'(classified as 'Low', 'Medium' or 'High') and 'Feeding period' (classified as Pre-feeding, Feeding or After feeding) and variable output is a Stress Index that was defined by using literature as S = 2*(Frequency of agressive behavior) + 1*(Frequency of steryotypic behavior) – 1*(Frequency of physical affect between family members). We chose Fuzzy Logic because this problem involves imprecision and Fuzzy Logic provides the development of algorithms, which are able to represent uncertainty inherent in data, allows the description, in linguistic terms, of problems that should be solved and can be an advantage in cases where an explicit analytical-process model is not available or is too complex (e.g. How much visitors represent the group 'High Visitor numbers'?). We have verified if the mathematical model can explain stress behavior patterns on brown capuchin at the Zoo Park Quinzinho de Barros in Sorocaba, Brazil by comparing data collected on field and observed model results. The social behavior of two brown capuchin (one adult male and one adult female with one baby brown capuchin on her back) was studied over 20 observation hours in November 2012 by using focal animal sampling. The difference between predicted (model results) and observed change (field data) in stress index over time were not significantly different by using Fisher's Exact test(N = time intervals = 90, $P>0.99$). The study revealed that fuzzy process is a tool to help on studies based on subjective information as brown capuchin behaviors.

Temperament of dairy calves subjected to in two different types of housing and management

Nathasha Radmila Freitas[1,2], Mateus José Rodrigues Paranhos Da Costa[2], Lívia Carolina Magalhães Silva[2,3], Aline Cristina Sant'anna[2,4], Luciana Pontes da Silva[2,3] and Maria Fernanda Martin do Amaral Guimarães[2,3]
[1]*Graduação em Ciências Biológicas, FCAV-UNESP, Via de Acesso Prof. Paulo Donato Castellane s/n, 14884-900, Jaboticabal, Brazil,* [2]*FCAV-UNESP, Grupo ETCO, Departamento de Zootecnia, Via de Acesso Prof. Paulo Donato Castellane s/n, 14884-900, Jaboticabal, Brazil,* [3]*Programa de Pós-Graduação em Zootecnia, FCAV-UNESP, Via de Acesso Prof. Paulo Donato Castellane s/n, 14884-900, Jaboticabal, Brazil,* [4]*Programa de Pós-Graduação em Genética e Melhoramento Animal, FCAV-UNESP, Via de Acesso Prof. Paulo Donato Castellane s/n, 14884-900, Jaboticabal, Brazil; natharfreitas@yahoo.com.br*

The improvement of handling conditions can amend the temperament of dairy calves and facilitate the labor routine at dairy farms. The aim of this study was to evaluate the effect of two different types of handling during the suckling period on the temperament of dairy calves. Two handlings strategies were compared: (1) conventional handling (CH), adopted in many dairy farms in Brazil, maintaining the calves tied (individually) in small shelters (from the 4th days of age until weaning,) calves were tied with a 2 m long chain and received milk direct in a bucket, receiving few interactions with the handler; (2) rational handling (RH), incorporating a set of good practice of handling, including group raising, suckling in buckets with nipples, frequent social interaction with the handler (including tactile stimulation) and progressive weaning. In both cases the calves were weaned when they reached 70 kg of life weight, around 70 days old. Twenty-three crossbreed Holstein and Jersey or Gir calves were evaluated two times: at 6 and 10 months old. Three indicators of temperament were used: flight distance (FD), novel object test (NO) and flight speed (FS). The data were analyzed using the t test for continuous measures (FD and FS) and Chi-Square test to the NO. There were no significant differences between the treatments for FS (RH=0.78±0.29 and CH=0.61±0.19 m/s, t=-1.69, $P>0.05$), FD (RH=1.75±0.81 and CH =1.87±0.81 m, t=0.45, $P>0.05$) and NO (χ^2=4.58, $P=0.33$). We concluded that the two evaluated managements (handling strategies, housing) did not influence the expression of calves' temperament. The conditions of the temperament tests (unfamiliar place, unusual handling) may influence the results. So, it is recommended to study add another test in calves' familiar place.

How dogs process familiar and inverted faces?

Sanni Somppi[1], Heini Törnqvist[1,2], Laura Hänninen[1], Christina Krause[2] and Outi Vainio[1]
[1]Research centre for animal welfare, P.O. Box 57, 00014 University of Helsinki, Finland, [2]Cognitive Science, P.O. Box 9, 00014 University of Helsinki, Finland; laura.hanninen@helsinki.fi

Face recognition plays an important role in social mammals. Facial-inversion effect and recognition of familiarity are widely studied phenomenons in primate face perception research but not studied before in domestic dogs with different life experiences. We tested with a contact-free eye-tracking system an untrained gazing behaviour of 8 kenneled purpose-bred laboratory beagles (4 y) and 23 private owned family dogs (4.5±3 y) for upright versus inverted pictures and for familiar versus strange dog and human faces. Prior to the experiments the participating dogs were trained with positive operant technique to lie still and lean their chin in a special designed rack. Each image was shown 1,500 ms and followed by a 500 ms blank gray screen. Images were digital photos of faces of familiar dogs (n=3), familiar humans (n=3), strange dogs (n=3) and strange humans (n=3). Strange faces were presented also in their inverted orientation. Each image was presented three to four times during the two-day-experiment (60 images/day). One trial consisted of 8-12 images and each dog was shown 6 trials. The length of the trials and the order of the stimuli were randomized and the dogs were rewarded after every trial. We recorded the number and total duration of fixations targeted to face and eye areas of the images. The effects of dog type (purpose-bred vs. family housed) and images (human vs dog, upright vs. inverted and familiar vs. strange) on gazing behaviours were analysed with repeated measure linear mixed models. Overall dogs fixated equally long on upright and inverted images but the eyes of the upright faces gathered longer fixations than inverted faces (111.1±27.2 ms, vs. 93.4±27.2 ms, $P=0.02$). Also, upright dog faces attracted fixations more often (1.3±0.3 vs. 1.2±0.2, $P=0.02$) and for longer duration than upright human faces (247.7±40.4 ms vs. 209.1±40.7 ms, $P=0.02$). Dogs fixated more often on familiar faces than strange faces (1.7±0.3 vs. 1.3±0.3, $P=0.02$). Respectively, eye areas of familiar faces gathered more fixations than of strange faces ($P=0.02$). Overall laboratory dogs fixated shorter duration to the face area than family dogs (145.3±68.8 ms vs. 311.7±40.4 ms, $P=0.05$). We concluded that the eye movements of dogs are guided by semantic information such as familiarity and direction. Although the studied two dog populations have very different social backgrounds and rearing environments, their gazing behaviours were quite similar. Eye movement tracking is a promising method for the broader exploration of processes underlying the cognitive skills of dogs.

Alternative bedding substrates during finishing affect lamb behaviour

Dayane L. Teixeira[1], Genaro Miranda-De La Lama[2], Morris Villarroel[3], Juan Escós[2] and Gustavo A. María[2]
[1]*Teagasc, Moorepark, Pig Development Department, Animal & Grassland Research & Innovation Centre, Fermoy, Co. Cork, Ireland,* [2]*University of Zaragoza, Faculty of Veterinary Medicine, Department of Animal Production and Food Science, Miguel Servet 177, 50013 Zaragoza, Spain,* [3]*ETSIA Polytechnic University of Madrid, Department of Animal Science, Avenida Complutense s/n, Ciudad Universitaria, 28040 Madrid, Spain; dayane.teixeira@teagasc.ie*

The study compared the effect of four types of bedding (sawdust, cellulose, straw and rice husk) on the behaviour of lambs (n=96 entire males, 19.11 ±1.07 kg live weight, approximately 80 days old) during the finishing phase of fattening (14 days) in an indoor system. Twelve lambs from each treatment were provided with only one type of bedding, fed with concentrate and water *ad libitum*. Two pens per treatment were evaluated. Live weight and concentrate consumption were recorded to calculate average daily gain and the conversion index and were analysed using GLM. Behaviour data were analysed by Kruskal-Wallis and Mann-Whitney-U tests using lambs as unit of measure. In general, lambs spent more time lying on straw (79.07%) and less on cellulose (75.25%), compared to sawdust (77.57%) and rice husk (76.71; $P \leq 0.05$). In contrast, lambs on sawdust (2.08%) and cellulose (1.92%) spent more time walking than lambs on straw (1.52%; $P \leq 0.05$). Eating behaviour was more frequent for lambs on straw (7.18%) than on sawdust (6.46%), cellulose (6.49%) and rice husk (6.05%; $P \leq 0.05$) but drinking behaviour did not differ among treatments. On week 1, the lambs on straw spent at least 2.43% more time lying than the others treatments ($P \leq 0.05$). By week 2, the lying frequency of lambs on cellulose was similar to the straw group ($P \leq 0.05$). Lambs on sawdust decreased 24.16% on their frequency of walking between week 1 and 2 ($P \leq 0.05$). Eating and drinking behaviours did not differ between weeks among treatments. No significant differences in productive performance or carcass quality traits were observed among treatments. All substrates evaluated could be efficiently incorporated into the system as bedding material without affecting productivity or product quality.

Welfare Quality – study, adaptation and application to the Brazilian dairy cattle

Guilherme Amorim Franchi[1], Paulo Rogério Garcia[2] and Iran José Oliveira Da Silva[2]
[1]Wageningen University and Research, Droevendaalsesteeg 2, 6708 PB Wageningen, the Netherlands, [2]Escola Superior de Agricultura Luiz de Queiróz – ESALQ/USP, Engenharia de Biossistemas, Avenida Pádua Dias 11, 13418-900 Piracicaba, Brazil; iranoliveira@usp.br

Due to the necessity of establishing animal welfare standards for the Brazilian dairy sector in harmony to the new consumer's requirements and legislation, it was drawn up the project Welfare Quality (WQ) - Brazil, based on the proposed project Welfare Quality ® European Union for dairy cattle. The assessments of animal welfare were performed in seven dairy farms of different technology levels in São Paulo/Brazil. They were selected in order to represent the main types of dairy farms found in Brazil, according to feeding and milking systems, breed selection, availability of outdoor area and milk production per cow, mainly. To carry out the project, it was used the evaluation protocol of welfare in Dairy Cattle Welfare Quality ® Assessment Protocol for Cattle, which is based on the principles of Good Feeding, Proper Installation, Good Health and Appropriate Behavior. The protocol defines four possible categories for the assessed dairy farms: Not classified, Acceptable, Good or Excellent. Only one farm received category 'Acceptable', because of a poor 'Good Feeding' score, while the others received category 'Good'. Furthermore, other highlight is the unsatisfactory score for the principle 'Appropriate Behavior' received by four farms. Possible reasons are inappropriate animals handling, assessor's subjectivity and/or protocol's subjectivity. To this final point, some emotion standards are vague and do not describe how animals should behave for each type of situation during evaluation. Finally, it can be concluded that the European protocol for the Evaluation of Welfare in Dairy Cattle Welfare Quality ® might be used in Brazilian dairy farms provided there is previous assessor training and adaptation of some measures and definitions to Brazilian dairy sector's reality, in which the outdoor dairy systems are the most predominant.

Ability to identify individuals similar in function of different quantities and color

Ana Carolina Donofre, Iran José Oliveira Da Silva and Fabiane Coletti Furlan
Escola Superior de Agricultura, Departamento de Engenharia de Biossistemas, Núcleo de Pesquisa em Ambiência, Avenida Pádua Dias, nº 11, 13418-900, Piracicaba, SP, Brazil; acdonofre@gmail.com

This study aimed to investigate the ability of newborn chicks to identify individuals similar in function of different quantities and colors. Twenty chicks were housed individually for 48 hours without none visual contact with each other. The treatments consisted of the presence (T1) or absence (T2) of a CS (social companion represented as a yellow ball, simulating a real bird) during 48 hours. Two preferences tests were conducted, both with four options. The first test had different amounts of CS (0, 1, 3 and 5), the other test had different colors CS (yellow, red, blue and black). It was observed the number of fold and the time of occupancy in each of the four choices. Analyzes were performed with the Kruskal-Wallis test ($P<0.05$). In the first test (quantities), the frequency of occupation differed significantly between T1 and T2, because birds of T2 (without the social companion), had a higher occupancy (24 times) with 3 option in CS. The occupation time was the option with the highest number of CS ($P<0.05$) for both treatments, totalizing an average of 108 seconds (T1) and 110 seconds (T2). In the second test (color), there were significant differences ($P<0.05$) among treatments, the frequency (23 times) and average occupancy time (103 seconds) to red, both results with higher in T2. Apparently the birds of T1 chased the CS yellow (same color used in the adaptation time). It can be concluded that both treatments sought compartments with higher CS, and not even counted the previous T2 with CS. Regarding the color, birds without contact with CS demonstrated to prefer red.

Assessment of human-animal relationship in broilers with automatic recording of activity

Anna M Johansson and Harry J Blokhuis
Swedish University of Agricultural Sciences, Animal Environment and Health, Box 7068, 75007 Uppsala, Sweden; anna.ma.johansson@aol.se

Broiler production consists of large flocks of thousands of animals. The farmer walks among the birds daily, but has in comparison with other livestock little physical contact with individual animals. Even with this limited contact, studies show that the relationship between the caretaker and the poultry flock can affect both the animals' welfare and the productivity. In this pilot study, two small groups of broilers were reared for 36 days. One of the groups was treated with additional human contact every day to make the birds more familiar to humans. Welfare assessment according to the Welfare Quality® protocol for poultry (2009) was carried out once a week, with special emphasis on human-animal relationship using the Avoidance Distance Test (ADT). Video recording of the birds' response to the presence of a human was carried out once a week in both groups. The recordings started with a human walking through the flock. Analysis of the activity of the birds after the human had left the pen was carried out using image analysis software. Activity was for each second expressed as an index between 0 and 1. Results show that the additional human contact group had higher values on the ADT-test than the control group: 91% of total number of birds could be touched compared to 65%. At week 5 the activity level during 20 minutes after a human had left the pen, was lower for the additional human contact group than for the control group (average index level 0.06 compared to 0.09). The results from this pilot study show a potential of using automatic recording of activity of broilers to assess the human-animal relationship but this needs further development.

Selection of dogs for assisted interventions, an exploratory study

Carolina Faria Pires Gama Rocha[1], Emma Otta[1], Marie Odile Monier Chelini[1], Katia Aiello[2] and Sara Favinha[3]
[1]*Psychology Institute, São Paulo University, Experimental Psychology Department, Av Prof Mello Moraes, 1721, Bloco C, Cidade Universitária, 05508-030, Brazil,* [2]*Instituto Nacional de Ações e Terapias Assistidas por Animais INATAA, São Paulo, SP, Brazil,* [3]*Terapias assistidas por Animais TAC, São Paulo, SP, Brazil; carolina.faria.rocha@gmail.com*

There is no official certification for dogs working in therapy and activity assisted interventions (AAI) in Brazil. Selection is grounded on loosely categorized intuitive observations based on handler expertise and brief organization guidelines. The purpose of this study is to evaluate the selection methods used by two of the biggest não governmental organizations (NGO 1 and 2) working with AAI in Brazil. To evaluate these methods 17 dogs (NGO1: n=10, NGO2: n=7) that have been approved by both NGO's selection protocol, were evaluated with internationally recognized behavior tests, the 2003 Canine Behavioral Assessment and Research Questionnaire-CBARQ and the 2012 Jakovcevic's sociability test. Both dog's groups were quite heterogeneous regarding breeds and training time (NGO1 12.1±20.2 month, NGO2 30.7±29.2 months). They were compared relatively to mean CBARQ classes and sociability results using SPSS for Windows software. T-test did not show any significant differences between groups but CBARQ results pointed out that 82% of the animals from both groups presented serious behavioral issues such as dog directed fear and aggression, and attachment/attention seeking. A positive trend was found by Pearson's correlation (r=0.452, $P=0.068$) between sociability test variable 'total time close to experimenter', which measured the time the dog stayed closer to 1 meter of the experimenter (NGO1: 96±76 sec, NGO2: 196±54 sec), and basic and animal assistance training history These results suggest a potential association between a continuous training and socialization, and an improved sociability indicator when interacting with a stranger person. In conclusion, our results indicate that the methodology used for dogs selection in these NGOs presents flaws and must be improved. Animals with behavior issues should be clearly identified. A more adequate and extensive training may be useful to solve some of the problems.

Behavioural and hormonal responses of timber wolves to training: implications for their husbandry

Angélica Silva Vasconcellos[1,2], Zsófia Virányi[1,3,4], Friederike Range[1,3,4], César Ades[2] and Kurt Kotrschal[1,3]
[1]*Wolf Science Center, Dörfles 48, 2115 Ernstbrunn, Austria,* [2]*Institute of Psychology, University of São Paulo, Experimental Psychology, Av. Professor Mello Moraes 1721, 05508-900 São Paulo, Brazil,* [3]*University of Vienna, Zoology Department, Althanstraße 14, 1090 Vienna, Austria,* [4]*Messerli Research Institute, Veterinärplatz 1, 1210 Vienna, Austria; angelicavasconcellos@gmail.com*

Wolves have been increasingly trained and socialised for studies in cognition and communication, processes which demand intensive contact with humans. Training has positive effects on dogs; however, its effect on Canis lupus, which may differ from domestic dogs in this respect due to the lack of domestication, has not yet been assessed. In order to examine responses of socialised timber wolves to training, we studied nine 1-2-year-old wolves (six males, three females). The animals had been hand-reared and were maintained for studies in cognition, trained from birth three times a week. In 15 five-minute sessions, the wolves were trained to attend to commands using positive reinforcement (cheese) in a training room, with five different trainers (randomly assigned), on a voluntary basis. We evaluated: percentage of time wolf oriented to trainer and spent close to he/she, mean latency to obey, percentage of commands correctly attended, and variations in wolves' saliva cortisol (as a measure of stress, taken before and 15 minutes after the sessions). Non-training related behaviours (NTRB: exploring, retreating, jumping) were observed, although at low rates (summing <10% of the time). Percentage of time spent within one meter of the trainers was 90.40±0.88%, and oriented towards them 85.95±1.01%. Percentage of correct commands was 65.48±1.19%, with a mean latency 1.16±0.08 seconds. Cortisol concentrations were higher in the second and third sessions, but reduced after every session, revealing also an effect of trainer and wolf identity (GLM, Cortisol $P=0.003$; Trainer $P=0.000$; Wolf $P=0.000$; Session $P=0.012$). The reduction in the wolves' cortisol concentrations after every session, their performance and low rates of NTRB indicate that training is not likely to cause distress in hand-reared timber wolves. The reason for the cortisol increase in the second session (maintained in the third session) is not clear, but may be due to an effect of anticipation.

Fear in Swedish farmed mink (*Neovison vison*)

Lina Olofsson, Anna-Maria Andersson and Per Wallgren
The National Veterinary Institute of Sweden, Ulls väg 2B, 751 89 Uppsala, Sweden;
dina90@hotmail.com

Minimizing fear in farm animals is essential for animal welfare and production issues. When 1,696 mink at 17 Swedish farms were tested by inserting a wooden spatula through the front of the cage, the immediate reaction in 83% of the mink was explorative (range: 63-94% at herd level). The percentage of truly fearful mink in Sweden was low (2%; range 0-12%) compared to previously published prevalence of fear in adult mink in Scandinavia (14-39%), but almost identical to the prevalence in Canada (2.5%). The interpretation of aggressiveness is not clear as mink showing aggressiveness in the applied 'stick test' shows confidence in the alternative 'glove test'. Therefore, aggressiveness (Mean 4%; range: 0-17%) was not evaluated. However, 9% of the mink acted hesitantly (range: 0-30%) and 1% remained in the nest box despite being alerted prior to the test (range 0-5%). Considering that fearful animals to a higher extent remain in the nest box on approach of a fearful stimulus and that 'freezing' could be a passive reaction of fear, these two behaviours may be regarded as potentially fearful. Thus, the prevalence of fearful mink could be as high as 12% (range: 2-34%). The wide range of truly, as well as potentially, fearful mink between farms reminds that this is a continuous work, and that further improvements in reducing fear could be made in certain farms. In conclusion, by refining the interpretation of the 'stick test' by including hesitant and in nest box behaviours, potentially fearful mink can be detected in low fear populations and this facilitates further progress in the reduction of fear in these populations.

Body temperature response to capture and restraint in migratory passerines at a stopover site

Fernanda M Tahamtani[1], Ivan Maggini[2,3], Massimiliano Cardinale[4], Leonida Fusani[2] and Claudio Carere[5]
[1]University of Edinburgh, Royal (Dick) School of Veterinary Studies, Edinburgh, EH25 9RG, United Kingdom, [2]University of Ferrara, Province of Ferrara, 44121, Italy, [3]University of Western Ontario, Advanced Facility for Avian Research, London ON, N6A 3K7, Canada, [4]Swedish University of Agricultural Sciences, Department of Aquatic Resources, Institute of Marine Reseach, Lysekil, 45330, Sweden, [5]University of Tuscia, Department of Ecological and Biological Sciences, 01100 Viterbo, Italy; fmtahamtani@gmail.com

Capturing wild birds with mistnets is a widely used procedure for studying several aspects of avian biology; however, little is known about physiological responses to such procedure. Here we focussed on body temperature (BT) response in migratory songbirds. We captured 63 garden warblers, 12 whinchats and 11 barn swallows at the Ponza ornithological station (40° 50' N, 12° 58' E), an island stopover site in central Mediterranean, during spring migration. BT was measured within three minutes (T0), 30 min (T30) and 180 min (T180) from capture using an analogue probe inserted into the birds' throat. A three hour restraint period is longer than the usual time needed for catching/ringing procedures (max. one hour when many birds are in the queue) but common for scientific experimentation e.g. on temporal dynamic of corticosteroid response (Silverin, Poul. Avian Biol. Rev 9:153, 1998). In between measurements birds were kept in cotton bags in a quiet shaded area and sat calmly inside the bags, only struggling when the experimenter went to catch them for measurements. The birds were then weighted and scored for a condition index and released. Body condition index (condition) was calculated via PCA of weight, and scores of the size of fat deposits and pectoral muscles. Mean DeltaT (calculated by subtracting T0 from T180) for garden warblers, whinchats and swallows were, respectively, -1.6±1.3, -0.46±2.05 and -0.49±1.52 with some birds decreasing temperature by up to 4.1 °C in garden warblers, 4.9 °C in whinchats and 3.0 °C in barn swallows. In whinchats lean individuals had lower body temperatures at T180 ($P=0.045$), and females tended to decrease BT, while males did not, across time ($P=0.020$). Garden warblers significantly decreased BT across time ($P<0.0001$) irrespective of body condition. Barn swallows presented an inverse relationship between DeltaT and condition with fatter birds displaying larger temperature drops in the three hour period ($P=0.016$). All birds flew away normally at release. We highlight potential welfare issues and suggestions concerning capture and handling of small birds, especially lean individuals, by mistnetting. In addition we discuss these results in light of the hypothesis of adaptive hypothermia during migration.

The effects of light and dark on lying behaviour, sleep, IGF-1 and serotonin in dairy cows

Emma Ternman[1], Sabine Ferneborg[1], Per Peetz Nielsen[1], Laura Hänninen[2], Ahmed Salama[3], Gerardo Caja[3] and Sigrid Agenäs[1]
[1]Swedish University of Agricultural Sciences, Kungsängen Research centre, 753 23 Uppsala, Sweden, [2]University of Helsinki, Research Centre for Animal Welfare, P.O. Box 57, 00014 University of Helsinki, Finland, [3]Universitat Autònoma de Barcelona, V0-354, Fac. Veterinaria, Campus de la UAB, 08193 Bellaterra, Barcelona, Spain; emma.ternman@slu.se

Knowledge of the effects of light on sleep and rest in livestock is scarce. We studied the effect of permanent light (24-0) and short day (4-20; 4 h light: 09:30 to 13:30, and 20 h dark) on sleep and rest behaviour, and photoperiod related hormones (IGF-1 and serotonin) of dairy cows. The study was approved by Uppsala local ethics committee. Five Swedish Red dairy cows in mid-lactation were randomly allocated in two unbalanced groups and submitted to a changeover design during the end of winter (March). Cows were housed in a windowless barn and adapted to light-dark treatments for 72 h, after which measurements lasted for 24 h each. Sleep and general activity (activity data obtained from four cows) were recorded by a portable EEG device and a pedometer fastened on the right hind leg. Data were split in circadian day- and night time periods (12 h) and analysed using pair-wise t-test (presented as means ± SE). Blood samples (10 ml) for hormones, were taken hourly through a permanent jugular cannula. Cows received sedation and analgesia before cannula was inserted, sampling started 24h after insertion. No complications were noted during or after the study. The data were analysed using procedure mixed models in SAS (presented as ls means ± SE). 24-0 treatment resulted in longer total REM sleep time ($P<0.05$) compared to 4-20 treatment (62±14 vs 39±10 min). Total standing time tended to be longer ($P<0.1$) during circadian night time compared to circadian daytime in 24-0 (428±70 vs 297±49 min). Total lying time was longer ($P<0.05$) during circadian night time in 4-20 compared to circadian night time in 24-0 (381±67 vs 293±70min). Plasma levels of IGF-1 (133±2 vs. 124±2 ng/ml; $P<0.01$) and serotonin (257±11 vs. 183±13 ng/ml; $P<0.05$) were higher in 4-20 compared to 24-0. This pilot study showed that permanent light may affect the distribution of rest and sleep. Furthermore, permanent light alter the hormone production necessary for maintaining biological rhythms.

Effect of an intermittent sound on the behaviour and heartbeat parameters of goats

Julia Johns[1], Antonia Patt[2] and Edna Hillmann[1]
[1]Animal Behavior, Health and Welfare Unit, Institute of Agricultural Sciences ETH Zurich, Universitaetstrasse 2, 8092 Zurich, Switzerland, [2]Centre for Proper Housing of Ruminants and Pigs, Federal Veterinary Office, Agroscope Reckenholz-Taenikon Research Station ART, Taenikon 1, 8356 Ettenhausen, Switzerland; julia.johns@usys.ethz.ch

Studies on the auditory capacities of goats showed that goats perceive sounds very similar to humans, and it is thus assumed that noise is as aversive to goats as it is to humans. However, systematic investigations on the reaction of goats to sounds of high amplitudes are missing. We investigated whether a chime of a bell with increasing amplitudes (41-96 dB) and a fundamental frequency of 1.9 kHz (4.7-14.3 kHz), affects feeding duration, duration of alertness, heart rate and heart rate variability. Twenty-seven goats were individually tested for a maximum of 5 minutes in a test arena on six consecutive days. In the test arena, goats were allowed to feed on hay. On the first day, reference values were recorded without playback for 5 minutes. On the following five days, the playback was conducted for one minute. Observations started immediately after the animal had entered the experimental room. The playback started when the goat was feeding for about 30 seconds and was terminated after one minute. The differences between test and reference values were analysed using linear mixed-effects models. With an increasing amplitude, the goats' heart rate variability decreased ($P=0.01$) indicating an increase in aversiveness with increasing amplitude. On the first and second day of playback, the feeding duration was shorter ($P<0.0001$) and goats were more alert ($P<0.0001$) than on the other days. However, goats seemed to habituate to the sound as heart rate ($P=0.0005$) decreased from the 1st to the 5th playback-day. Although the physiological response was not very pronounced, the results indicate that after a negative initial reaction, the goats habituated to the sound when exposed repeatedly.

Colour types and numbers of feeding places affect bite marks in group housed mink

Steffen Hansen and Jens Malmkvist
Aarhus University, Animal Science, Blichers Allé 20, P.O. Box 50, 8830 Tjele, Denmark;
Steffenw.hansen@agrsci.dk

Bite marks on the leather side of the fur can be quantified at pelting and thus represent an objective measure of the social interaction among mink kept in the same cage. Bite marks appear in group housed mink more often than in mink kept traditionally in pairs and thereby represent an essential criticism of group housing of mink. Therefore, we examined whether the number of feeding places and the amount of feed can reduce the occurrence of bite marks in mink kept in groups. The study included 192 mink kits placed in 48 climbing cages from weaning to pelting at the age of 7 months. In each cage were inserted 2 males and 2 females all with different colours. The study showed that access to 3 feeding places reduced the number of bite marks in the neck and body ($P<0.01$) and tail ($P<0.05$) compared to mink having access to one feeding place. A 10% reduction in the amount of feed did not affect the skin length or the number of bite marks. Black mink had more bite marks than Brown mink ($P<0.001$) and Brown mink had more bite marks than Palomino mink ($P<0.01$). Black and Brown female mink had more bite marks than similar males ($P<0.001$) whereas no differences could be shown between sex in light coloured mink. A reduced number of bite marks in light coloured mink are consistent with the hypothesis that bite marks are due to mechanical damages of the hair follicles and therefore more easily observed in dark mink with dark melanin granules than in light coloured mink without dark melanin granules.

Examination of *Salmonella* spp. and *E. coli* spp. in cage and cage free environments for laying hens

Dana Didde, Sheila Purdum and Kathryn Hanford
University of Nebraska-Lincoln, Animal Science, 3600 Fair St. Animal Science C206, Lincoln, NE 68583, USA; dhahn6@huskers.unl.edu

The implications of alternative housing of laying hens and food safety are important issues in the U.S. egg industry. A study was conducted to evaluate the effect of layer housing on environmental and egg shell counts for *Salmonella* and *E. coli* spp. Bovan Laying Hens (n=96) from 72-78 wks of age were assigned to 1 housing trts, traditional cages or cage free. Hens were housed in separate but identical rooms, with 4 cages and 4 cage free in each room; trts were blocked by room. There were 4 hens/cage (75 sq. in./hen); and 8 hens/floor pen, with 1.5 sq. ft./hen and 2 nest boxes. Cage or floor pen served as experimental unit. Measurements included daily egg production, weekly feed intake and egg wt, biweekly environmental and egg shell crush for *Salmonella* and *E. coli* testing. Environmental samples were aseptically collected, weighed and placed in a 1:10 solution. Egg shell crush was conducted by aseptically separating egg contents from the shell and crushing the shell in a 50 ml tube and placing it in a 1:10 dilution. There was a significant housing by time interaction ($P \leq 0.0005$) with cage hens having higher egg production than cage free hens. Feed intake was significant between cage and cage free ($P \leq 0.0006$), cage free requiring 14.6 g/hen/d more than cage hens. There was a significant housing by time interaction for egg wt ($P \leq 0.001$), there were no differences in egg mass (g feed:g egg) ($P \leq 0.565$). There were no differences in *Salmonella* prevalence in environmental samples or shell crush samples in cage and floor pen hens. There was a housing by time interaction for *E. coli* (≤ 0.001), cage free had lower CFU counts than cages. Thus, hens in cage free had equal prevalence of *Salmonella* as traditional cages, along with lower CFU counts for *E. coli*.

Reducing R/R-behavior in captive orangutans through dietary treatments: a case study

Paul Koene[1], Jan-Willem Haeke[1] and José Kok[2]
[1]*Wageningen University and Research Center, Livestock Research, De Elst 1, 6708 WD Wageningen, the Netherlands,* [2]*Ouwehand Zoo, Grebbeberg 111, 3911AV Rhenen, the Netherlands; paul.koene@wur.nl*

A group of eight orangutans (*Pongo pygmaeus*) is kept at Ouwehand Zoo. One particular individual, Tjintah, performs regurgitation and reingestion or R/R behavior, which is often found in primates. This behavior is defined as a self-induced movement of food or liquid from the esophagus or stomach to the floor, the hands or retained in the mouth, subsequently followed by reingestion. It is still unclear why this behavior occurs and how it is triggered. One of the potential causes is the difference between wild fruits and domestic fruits. Domestic fruits lack fibre and contain three times more sugars. This study aimed to reduce R/R in Tjintah through dietary treatments. The standard zoo diet was used as the control treatment (A) and compared with a newly formed vegetable diet (C: high fibre, low in sugar). In two additional treatments the standard fruit and the vegetable diet were fed every hour (B and D). The fifth treatment was a vegetable diet with additional browse (E). All diets fulfilled nutritional demands from the SSP Husbandry Manual and were fully consumed. In a randomized alternating treatment design (RATD), treatments were presented for 35 days with every day ten observations of 30 minutes focal continuous recording. During this period the average R/R frequency was 6.88±8.58 per 30 minutes. Single case randomization tests (SCRT) showed that the dietary treatment had a significant effect on the occurrence of R/R (SCRT, RATD, $P=0.002$). R/R frequency was highest when Tjintah was provided fruit diets (A=11.94, B=10.89) compared to the vegetable diets (C=4.94, D=4.30 and E=4.34; SCRT, RATD, $P=0.001$). This study showed that removing fruit from the orang-utan diet and replacing it with vegetables resulted in a reduction of R/R behaviour. Whether lower sugar or increased fibre content has caused this decline should be studied in future.

Influence of decoy posture on preference and investigative behaviour in Sika deer
Ken-ichi Takeda and Ryuji Yamazaki
Shinshu University, Faculty of Agriculture, 8304 Minamiminowa, 399-4598, Nagano, Japan;
ktakeda@shinshu-u.ac.jp

An overabundance of Sika deer (*Cervus nippon*; hereafter, deer) has caused dramatic increases in agricultural and forestry damage, and has severely affected natural vegetation and ecosystems in national parks and nature reserves by heavy foraging pressure. The deer population needs to be controlled, but the number of smart deer that are wary of people has increased, and the number of hunters who control the deer population has decreased dramatically in Japan. Therefore, an efficient method to attract and capture deer needs to be developed. We investigated the influence of the posture of life-sized decoys on deer preference and investigated deer behaviours. Six farmed hinds, have been reared in consideration of animal welfare, were used in individual preference tests. Each deer participated in one trial per day, and a total of seven trials. A deer was introduced to a starting box and presented with three different postured decoys at the same time. Subsequently, the deer was released to a preference test area where it could simultaneously distinguish three postured decoys (standing, lying and grazing posture) that were 10 m away. The first decoy that was chosen by a deer and all investigative behaviours by the deer towards any of the decoys during 30 min of continuous observation were recorded. There was no significant difference in deer preference for decoy location ($P=0.29$), and no location bias was found. Deer showed a significant preference for the grazing posture decoy ($59.5\pm10.6\%$) over the other decoys (Tukey-Kramer test, $P<0.01$). Deer preferred the lying posture decoy ($28.6\pm15.6\%$) over the standing posture decoy ($11.9\pm14.0\%$; $P<0.01$). Investigative behaviours by deer did not differ among postures, and the percentage of investigation for each part of the decoy was higher than the expected value (Tukey-Kramer test, $P<0.01$). In conclusion, grazing posture of decoy to attract deer is better than others.

Play behaviour and used space of dairy and beef calves living in a semi-natural environment

Robert Somers[1], Stella Huertas[2], Elena De Torres[2], Nirita Brand[1] and Frank Van Eerdenburg[1]
[1]Faculty of Veterinary Sciences, Utrecht University, Department of Farm Animal Health, Yalelaan 7, 3584 CL Utrecht, the Netherlands, [2]Facultad de Veterinaria, Universidad de la República, Instituto de Biociencias Veterinarias, Lasplaces 1550, 11600 Montevideo, Uruguay; stellamaris32@hotmail.com

The aim of this study was to examine the quantity, types and space requirements of play behaviour of calves living 24 hours a day on pasture during two months, June & July of 2012 in Uruguay, South America, with a range of minimum and maximum temperatures between 7 and 15 °C respectively. Observations were made on a daily basis, for 30 days, in 2 different herds. Herd A consisted of 23 beef calves (age 1-2 months) living together with 80 female adult cows on 70 ha of grassland. Herd B consisted of 18 dairy calves (2 months old), living without dams on a 1 ha pasture. The observations were visual. All play behavioural elements were classified as being a jump, buck, turn, run, head butt, body butt or mount and an ethogram was developed. In total, 625 elements of locomotor play and 138 of social play were seen. The average number of calves playing was 1.44, the average group size (of calves whether they were playing or not) was 3.79. Running was observed the most often (458 times), followed by head butting (100 times), bucking (81), body butting (76), jumps (64), turns (48) and lastly mounts (17). Jumps had a mean height of 1.26 times withers height, bucks 0.57 and mounts 1.39. The mean running distance was 13.5 m; play fighting required an average of 3.50 m^2. A high percentage of play involved multiple calves playing together (33.9%). In the present, descriptive, study that included only one group with and one group without a dam present, the group with the dam present had a lower prevalence of play behaviour.

Dairy cow feeding and grazing behaviour of tanniferous forages in relation to harvest time

Danielle Lombardi[1], Elsa Vasseur[1], Robert Berthiaume[2], Trevor Devries[3] and Renée Bergeron[1]
[1]Alfred Organic Dairy Research Center, University of Guelph, Campus d'Alfred, Alfred, K0B 1A0, Ontario, Canada, [2]Agriculture and Agri-food Canada, Lennoxville, J1M 0C8, Québec, Canada, [3]University of Guelph, Kemptville Campus, Kemptville, K0G 1J0, Ontario, Canada; evasseur@alfredc.uoguelph.ca

Tanniferous forages are rich in condensed tannins known for their anthelmintic properties but also for their bitter taste. Forage total non- structural carbohydrates (TNC) could mitigate bitterness because of their high palatability. Forage TNC's have been shown to vary with harvest time. The objective of this study was to quantify voluntary intake (Exp1) and preference (Exp2) of dairy cows for tanniferous forages (fresh and fermented) harvested in the morning and evening, and determine grazing preference (Exp3). Forages were chicory (*Chichorium intybus*) harvested at 07:00 h (CHICAM) and 18:00 h (CHICPM), birdsfoot trefoil (*Lotus corniculatus*) harvested at 0700 h (BIRDAM) and 18:00 h (BIRDPM), birdsfoot trefoil haylage harvested at 07:00 (BIHAYAM) and 18:00 (BIHAYPM), and alfalfa silage (CONTROL). All cows were subjected to *ad libitum* access for 30 min to the six novel forages to associate postingestive feedback and taste of each forage (adaptation period). In Exp1, single forages were offered *ad libitum* for 30 min to 14 cows to determine voluntary intake in a replicated 7×7 Latin square. In Exp2, every possible pair of forages (21 pairs) was presented for 30 min to 8 cows. Weight of forage before and after consumption was recorded for each cow and analyzed with the MIXED (Exp1) and MDS procedures of SAS (Exp2). In Exp3, grazing preference tests (12 cows) consisted in one day of forced choices (1 h) on each of the two pastures (chicory and birdsfoot), and two days of free choices (1 h) between the two pastures. Grazing time and cow location on pasture was assessed through 2-min scan sampling and analyzed with the UNIVARIATE procedure of SAS. Exp1. Fresh forages (CHICAM, CHICPM, BIRDAM, and BIRDPM) had the lowest voluntary intakes regardless of harvest time (consumed 2.3-3.6 times less than fermented forages on DM basis, $P<0.001$).Exp2. While offered in pairs, BIHAYPM was preferentially consumed over all other forages followed by CONTROL and BIHAYAM. No differences were found in preference among fresh forages regardless of harvest time. Exp3. When having choice between both fields, cows spent 71% of grazing time on birdsfoot and 23% on chicory ($t=-40$, $P=0.2$). However, when forced to remain on one pasture, grazing times did not vary. These results suggested that tanniferous forages have the potential to be used as alternative forages for dairy cows.

Behaviour of pasture-housed Holstein dairy cows approaching parturition

Christa A. Kurman and Peter D. Krawczel
The University of Tennessee, Department of Animal Science, 258 Brehm Animal Sciences Building, 2506 River Drive, Knoxville, TN 37996-4574, USA; pkrawcze@utk.edu

The late stages of gestation represent a challenging period for a dairy cow as she undergoes changes in diet, physiology, and social environment. Increasing cow comfort, which can be assessed by lying behaviour, may reduce some difficulties associated with late pregnancy and calving. One management option for increasing cow comfort during this stage of the lactation cycle is pasture-based housing as it lacks the constraints of confinement housing. The objective of our study was to quantify the lying behaviour of Holstein dairy cows housed on pasture during the 8 d before calving. Eighteen multiparous Holstein dairy cows were moved to pasture approximately 3 wk before their projected calving date. Lying time (hr/d), number of lying bouts (n/d), and lying bout duration (min/bout) were recorded at 1-min intervals over the 8 d before calving using a datalogger. Data analysis was performed using a GLM model in SAS with the cow as the experimental unit. Mean lying time was 11.0±0.4 h/d from d -8 to d -2 and no day differed from d -8 ($P>0.2$). Lying time decreased (8.8±0.6 h/d) on d -1 relative to d -8 (10.8±0.4 h/d, $P=0.003$). Similarly, lying bouts only differed on d -1 (5.9±0.9 bouts/d) relative to d -8 (8.9±0.6 bouts/d; $P=0.01$). Lying bout duration did not differ throughout the 8 d before parturition ($P>0.27$). Although lying time decreased during the day leading up to calving, the decreased in lying bouts may indicate a reduction of restless and greater comfort for dairy cows calving on pasture. A direct comparison of the behaviour of confinement-housed and pasture-housed dairy cows approaching parturition is needed to establish recommendations to promote cow welfare during this critical time.

Diurnal behavior of cattle from different genetic groups during winter and summer seasons

Andrea Ribeiro[1], Mauricio Alencar[2], Ana Luisa Paçó[1] and Adriana Ibelli[3]
[1]FMU, R. Ministro Nelson Hungria, 541, 05690-050, Brazil, [2]Embrapa CPPSE, Rod. Washington Luiz, 234, 13560-970, Brazil, [3]Embrapa CNPSA, CP 21, 89700-000, Brazil; andrearbr@yahoo.com.br

The Brazilian beef cattle production system is increasing the use of taurine breeds (*Bos taurus*), both adapted and non-adapted, in crossbred programs with zebu (*Bos indicus*) heifers. Therefore, there is a lack of information about the behavior and adaptation of these genetic groups (GG) in grazing systems of subtropical areas. The aim of this study is evaluate the behavior activity of Nellore (0% B. taurus and 100% adapted – NX), Senepol × Nellore (50% *B. taurus* and 75% adapted – SN) and Angus × Nellore (50% *B. taurus* and 50% adapted – AN) cattle during summer and winter seasons, in the Southeast of Brazil. The experiment was conducted at the Southeast – Embrapa Cattle Center (CPPSE) (22°01"S and 47°53"W), in São Carlos, Brazil. The behavior of six heifers, from each GG, managed together in Tanzania grass (*Panicum maximum*, J.) pasture, was evaluated during three days, from 7:00 a.m. to 6:00 p.m., in two consecutive years, during the summer and winter seasons, totalizing 12 heifers observed per GG. The pasture was managed in a rotational system with three days paddock occupation and the observations were conducted always in the second day. Grazing, ruminating and idling behavior and in other activities were recorded by visual direct observation, at a ten minutes intervals. Mean diurnal air temperature during winter and summer observations were 19.09 °C and 30.46 °C, respectively. The data were analyzed by the least squares method. There were significant effects of GG for all variables studied ($P<0.05$). AN and SN heifers grazed longer than NX and these remained longer idling. AN heifers stayed longer in other activities and ruminating but did not differ from NX ($P>0.05$). Independently of GG, grazing, ruminating, idling and other activities occupy about 56.6%; 19.6%; 16.5% and 5.0% of diurnal activities, respectively, during summer and 68.1%; 12.3%; 15.1% and 4.1% of diurnal activities, respectively, during winter. All three GG grazed less and ruminated longer during summer (373.8±5.5 min and 129.7±4.4 min, respectively) as compared to winter (449.9±5.5 min and 78.5±4.4 min, respectively) ($P<0.05$) and there were no differences among them during summer season ($P>0.05$). However, for the winter time AN heifers grazed longer (481.4±9.9 min) than NX (403.0±8.8 min) ($P<0.05$) and did not differ from SN (463.5±9.9 min) ($P>0.05$). NX ruminated (98.6±6.6 min) and remained in other activities (36.6±3.3min) longer ($P<0.05$) than SN (60.3±6.6min, 29.8±3.3min, respectively) and AN (76.5±6.6min, 15.7±3.3 min, respectively). Based on these results, diurnal behaviour varies according to GG and season of the year.

Behavioural indicators of cattle welfare in silvopastoral systems in the tropics of México

Lucía Améndola[1], Francisco Javier Solorio[2], Carlos González-Rebeles[1] and Francisco Galindo[1]
[1]Universidad Nacional Autónoma de México, Etología, Fauna Silvestre y Animales de Laboratorio, Ciudad Universitaria, Circuito Exterior, Coyoacán, 04510 México, D.F., Mexico, [2]Universidad Autónoma de Yucatán, Facultad de Medicina Veterinaria y Zootecnia, Apdo. 4-116 Itzimná, 97100, Mérida Yucatán, Mexico; luciamendola@gmail.com

Animal production has been associated to several detrimental effects on the environment, but it is fundamental for achieving food security. Silvopastoral systems have shown to be a good alternative for sustainable livestock production that may provide ecosystem services and improve animal welfare. Therefore, the aim of the study was to compare the welfare of cattle in an intensive silvopastoral system based on high densities of leucaena shrubs (*Leucaena leucocephala*) combined with two tipes of grasses, guinea grass (*Panicum maximum*) and star grass (*Cynodon nlemfuensis*), and in a monoculture system based on star grass, both systems in the Yucatan peninsula. We selected eight heifers per system, which were observed during four consecutive days from 7:20-15:30, in each of three paddocks per system, and in the dry and the rainy season. Measures of forage availavility, and daily heat and humidity, were also taken. GLMM models, that allow the inclusion of fixed factors and random factors with repeated measures of the same individuals, were carried out to compare the frequency of social interactions, as well as time budgets (i.e. grazing, resting, and ruminating) in both the silvopastoral system and the monoculture paddocks. Heifers in the monoculture system displayed 33% less non-agonistic interactions ($P \leq 0.05$), and only in the rainy season, showed 44% less agonistic behaviours ($P \leq 0.02$). The average resting time was 44% longer in the silvopastoral system ($P \leq 0.001$), and only in the monoculture system the foraging times were affected by temperature ($P \leq 0.001$) and humidity ($P \leq 0.001$). Our results indicate that heifers in silvopastoral systems may maintain more stable social bonds, benefit from better quality resting bouts and be less influenced by heat and humidity, suggesting that animal welfare was enhanced.

Effect of solar radiation on use of shade and behavior of Holstein cows in a tropical environment

Steffan Edward Octávio Oliveira, Alex Sandro Campos Maia, Cíntia Carol Melo Costa and Marcos Davi De Carvalho
São Paulo State University, Animal Science, Professor Paulo Donato Castellani Access route w/n, 14883-900, Brazil; liv_forrow@hotmail.com

The aim of this study was to verify if the behavior and use of shade by Holstein cows are influenced for solar radiation. Ten cows divided into five groups of two animals were used, then they were distributed in five treatments, being one for paddock. The treatments consisted of artificial shades with different blockage to radiation (T30=30%, T50= 50%, T70= 70%, T100= 100% of blockage and T0= without shade). All groups were evaluated in each treatments for five days. The following behaviors evaluated were time in the shade, time standing, time idle and time grazing. The observational period was divided in four classes of hours (1=8-10 h; 2=10-12 h; 3=12-14 h; 4=14-17 h). The meteorological and behavioral data were recorded from 8:00 to 17:00 in 15 minute intervals. Data were subjected to analysis of variance using the least squares method, and the adjusted means were compared by Tukey test ($P<0.05$). The animals used more shade in treatments T70 and T100 ($20.01\pm1.87\%$ and $23.23\pm1.87\%$) and on class of hour 2 and 3 ($15.42\pm1.67\%$ and $22.66\pm1.67\%$). In this period the solar radiation were higher (642.01 ± 3.36 W/m^2 and 668.11 ± 3.36 W/m^2), differing than other classes of hours ($P<0.05$). Also during this same classes of hours the animals spent lower percentage of time standing ($59.86\pm1.31\%$ and $59.99\pm1.31\%$) and spent more time idle ($48.61\pm1.12\%$ and $48.26\pm1.12\%$), in order to diminish the heat gain by conduction and metabolism. The time spent grazing was higher on class 4 ($21.67\pm0.87\%$) where the radiation and temperature were lower (369.12 ± 3.36 W/m^2 and 30.81 ± 0.067 °C), compared to the classes 2 and 3 ($P<0.05$) The direct solar radiation influenced the behavior of Holstein cows. Once the levels of radiation were higher they spent more time on shade, lying down and idle in shades with higher level of blockage.

Effect of pasture cutting interval time on grazing behaviour of dairy cows

Willian Goldoni Costa, Vitor Borghezan Mozerle, Jéssica Rocha Medeiros, Jéssica Sebold May, Fabiellen Cristina Pereira and Luiz Carlos Pinheiro Machado Filho
Universidade Federal de Santa Catarina, Rodovia Admar Gonzaga, 1346, Itacorubi, 88034-000 Florianópolis, SC, Brazil; pinheiro@cca.ufsc.br

The aim of this study was to evaluate the effect of varying pasture rotation times with grazing management – 28 d (T28), and 56 d (T56) rotation – on bite rate, grazing and rumination time in an intensive grazing system. Six groups of three lactating Jersey cows balanced for parity and lactating stage were used in a duplicated 3×3 Latin square design. The composition of the pasture was *Lolium multiforum, Avena sativa, Penissetum purpureum, Axonopus compressus, Trifolium repens* and *Vicia sativa*. The size of the pasture was 100 m^2 per cow per day. Live observations were performed four hours after morning milking twice per group per treatment on the fourth and fifth day after a three-day habituation period. All variables were statistically analysed by ANOVA using the MIXED procedure in SAS. Grazing time and bite rate were higher during the first two hours of observation after milking ($P<0.001$) compared to the last 2h, in all treatments. Comparing T28 and T56, cows in T28 cows spent more time grazing (21.27 and 18.89 events, respectively; $P=0.09$) and had higher bite rates (53.55 vs 47.35 bites per minute, respectively; $P<0.01$). Cows spent more time ruminating ($P<0.05$) in T56, 4.08 events, compared to T28, 1.95 events. Twenty-eight days of pasture rotation appeared to result in the most efficient grazing by having a higher bite rate and longer grazing time.

Cattle grazing behavior regarding the distance of dung and infections by gastrointestinal nematodes

Hizumi Lua Sarti Seó, Luiz Carlos Pinheiro Machado Filho, Luciana Honorato, Patrizia Ana Bricarello, Vitor Borghezan Mozerle and Ícaro Nóbrega
LETA – Lab. de Etologia Aplicada e B-E Animal, Universidade Federal de Santa Catarina, Zootecnia e Desenvolvimento Rural, CCA, Rod. Admar Gonzaga, 1346, Itacorubi, Florianópolis, SC, 88.034-001, Brazil; zoelua@hotmail.com

The behaviour that promotes avoidance of parasitical contamination is adaptive and associated with host immunity. This study aims to evaluate whether cattle gastrointestinal verminosis is related to the grazing distance of this species dung in naturally contaminated pasture. Based on the average faecal egg counts (FEC) in individual samples, 18 bovines, male, 18 months old, were allocated to 3 groups, according to many FEC observations: High infection: FEC>315; Intermediate infection: FEC 130-160; Low infection: FEC 40-70. All animals were treated with anthelmintic reducing FEC to 0 in order to analyze the new nematode infection challenge. Three observers collected grazing behaviour data simultaneously during 2.5 hours/week (instantaneous of 5 minutes), in a total of 12 weeks. The exact grazing distance was recorded whenever it was less than 1 m from feces. Fecal samples were collected for FEC fortnightly. The group with high infection grazed farther from feces (average of 50 cm) than the intermediate group (average of 41 cm, $P=0.02$) and the low group (average of 43 cm, $P=0.05$). FEC variations were observed throughout the experiment. The groups with low and moderate infections kept their FEC from 0 to 200 (average FEC=17) and the FEC in the group of high infection remained significantly higher, from 0 to 1000 (average of 128). These findings indicate that the most parasitized animals adjust their grazing behaviour due to their infection condition.

Productive, phisiological and social behaviour characterization of Holstein steers production system

Oscar Blumetto[1], Andrea Ruggia[1], Antonio Torres Salvador[2] and Arantxa Villagrá García[3]
[1]INIA, National Program of Meat and Wool Production, INIA Las Brujas Experimental Station, Ruta 48 km 10, 90200 Rincón del Colorado, Canelones, Uruguay, [2]UPV, 3Institute of Animal Science and Technology, Camino de Vera 14, 46022, Valencia, Spain, [3]IVIA, Animal Technology Center CITA, Polígono La Esperanza 100, 12400, Segorbe, Castellón, Spain; oblumetto@inia.org.uy

The objective of this study was to characterize three different production systems for Holstein steers through their effect on productive traits, physiological indicators, and social interaction. 48 Holstein castrated males of 16 weeks of age, were randomly divided into three groups, corresponding to three treatments: (T1) confined into a 210 square meters yard, (T2) confined into a similar yard with six hours of access to grassland, (T3) permanent placed at grassland. Animals were weighted every 14 days. Social behaviour was continuously registered, during twelve hours a day (7:00 to 19:00), three days per week, in four weeks distributed throughout 19 weeks of experiment. Blood samples were taken for cortisol and biochemical profile analysis. Live weight was analyzed using PROC MIXED with repeated measures within animals. Biochemical profile and cortisol values were logarithmic transformed (LNt), and analyzed by PROC GLM. Interactions between animals were expressed as a count, LNt and analyzed by PROC MIXED. Average daily gain (ADG) was 0.756 ± 0.829, 0.757 ± 0.676 and 0.730 ± 0.762 kg/day for T1, T2 and T3 respectively ($P=0.1254$). Respect to social behaviour, non-agonistic interactions did not differ significantly between treatments ($P=0.1496$) whereas agonistic interactions resulted higher ($P<0.0001$) in T1 (13.4 ± 13.6) than in T2 and T3 (7.4 ± 5.4 and 5.9 ± 4.6 interaction per day respectively). Cortisol concentration did not show statistically significant differences ($P=0.7189$), and means for T1, T2 and T3 were 2.15 ± 1.69, 2.54 ± 1.54 and 2.05 ± 0.81 µg/dl, respectively. For biochemical profile of blood serum, average values were inside the reference ranges except for CK and glucose, which levels exceeded them in the three treatments. Even though values were within the reference range, certain parameters presented significant differences between treatments and animals in T3 reached the highest concentration of urea 11.8 ± 0.86 ($P<0.001$) as well as the lowest value of Alkaline Phosphatase 67.9 ± 11.10 ($P=0.005$). There were no biochemical evidences of increasing stress or health problems in any production system. However, permanent confined animals increased agonistic behaviour, which could reflects some welfare problems.

Deleterious effect of social instability on meat-type chicken learning abilities and behaviour

Carole Foucher[1], Ludovic Calandreau[1], Aline Bertin[1], Laure Bignon[2], Sarah Guardia[2], Elisabeth Le Bihan-Duval[3], Cécile Berri[3] and Cécile Arnould[1]
[1]INRA, UMR85 Physiologie de la Reproduction et des Comportements; CNRS UMR7247; Université de Tours; IFCE, 37380 Nouzilly, France, [2]ITAVI, Centre Val de Loire, 37380 Nouzilly, France, [3]INRA, UR83 Recherches Avicoles, 37380 Nouzilly, France; cecile.arnould@tours.inra.fr

Meat-type chickens are reared in very large groups and are submitted to repeated encounters with unknown conspecifics. Our aim was to assess the consequences of these encounters on intermediate 'certified' meat-type chicken (crossing of a fast-growing sire and a slow-growing dam) learning abilities, social behaviour and emotional reactivity to unknown situations. These chickens are likely to be more affected by these encounters than fast-growing ones. Indeed, they are slaughtered later and thus are sexually more mature at the end of the rearing period. Groups of 5 meat-type chickens were reared under stable social conditions (n=16 groups) or under unstable ones (i.e. the 5 birds of the groups changed regularly, n=16 groups) from Day 0 (D0) to D51. One chicken per group was tested in an associative learning task between D13 and D17. A very attractive feed (meal worms) was delivered in a particular environment with coloured strips and preference for this environment was then tested. Social behaviours were analyzed from D21-43. Bird emotional reactivity when faced to an unknown situation was assessed at the end of the rearing period and wing flapping on the shackle line during slaughtering (D46-51). Only birds from the stable condition were able to associate the attractive feed to the environment it was delivered ($P=0.01$, post-hoc test after ANOVA). Furthermore, these birds had a higher social proximity than those from the unstable condition ($P<0.01$, Mann-Whitney U test) and showed less aggressive interactions such as aggressive pecking ($P=0.06$), threats ($P<0.01$) or wing flapping in front of another bird ($P=0.03$). The social conditions tested had no effect on their emotional reactivity or reactivity during slaughtering in our testing conditions. These results underline that social instability increase aggressive interactions between conspecifics, but also have negative effects on some learning abilities which could have deleterious consequences on bird adaptation to their rearing environment.

Dominance behaviour of equines while being supplemented with concentrate in individual troughs

Marina Pagliai Ferreira Da Luz[1], Jose Nicolau Puoli Filho[1], Marcos Chiquitelli Neto[2], Cesar Rodrigo Surian[1] and Heraldo Cesar Gonçalves[1]
[1]*School of Veterinary Medicine and Animal Science, FMVZ, UNESP, Animal Production, Distrito de Rubião Junior s/n, 18618-970 Botucatu, SP, Brazil,* [2]*School of Engineering and Animal Science, FEIS, UNESP, Ilha Solteria, SP, Biologic and Animal Science, Avenida Brasil, 56 Bairro: Centro, 15385-000 Ilha Solteira, SP, Brazil; marina_pagliai@hotmail.com*

The equine hierarchy and the way it presents itself, during feeding time, is quite important to understand the dominance behavior. The objective of the present work, through visual observation, was to evaluate if the ratio between number of troughs and number of horses (rc/c=1 or 1,5), the distance from the ground, of the troughs (low=0 or high=45 cm) and the distance (d=1,6 m or 10 m) among them, affected the variables related to aggressiveness (pushing with head / neck, kicking, biting, pinned ear) of 8 male equines, of 380 kg and 8 years of age, respectively. The horses were supplemented daily with 2 kg of concentrate in individual troughs in 8 treatments: 1. high, d=1,6 m, rc/c =1; 2. high, d=1,6 m, rc/c 1,5; 3. high, d=10 m, rc/c=1; 4. high, d=10 m, rc/c=1,5; 5. low, d=1,6 m, rc/c=1; 6. low, d=1,6 m, rc/c=1,5; 7. low, d =10 m, rc/c=1; 8. low, d=10 m, rc/c=1,5. The data were collected every two minutes by a single observer. There was no significant statistical difference, according Kruskal-Wallis Test ($P<0,05$) for behaviors like biting and kicking. However, there was difference ($P<0,05$) between treatment 1 and 8 for pinned ears, pushing with head / neck and between treatment 6 and 8 for pinned ears. The conclusion is that the trough at the ground level, greater distance and ratio between trough/horses decreased pushing with the neck and pinned ears. The study was approved by the Animal Research Ethical Committee under the protocol # 147/2012.

From loneliness to a social life: the case of a chimpanzee at the national zoo of Chile

Alexandra Guerra[1], Guillermo Cubillos[2], Beatriz Zapata[1,3], Andrea Caiozzi[1,3,4] and Marcial Beltrami[1,5]

[1]*Universidad Mayor, Escuela de Medicina Veterinaria, Camino La Pirámide 5750, 8580000 Huechuraba, Chile,* [2]*Zoológico Nacional del Parque Metropolitano de Santiago, Sección Manejo y Bienestar Animal, Pío Nono 450, 8420000 Recoleta, Chile,* [3]*Universidad Mayor, Unidad de Etología y Bienestar Animal (UnEBA), Camino La Pirámide 5750, 8580000 Huechuraba, Chile,* [4]*Asociación Latinoamericana de Parques Zoológicos y Acuarios, Avenida Cristóbal Colón 4840, departamento 124-A, 7550000 Las Condes, Chile,* [5]*Universidad Metropolitana de Ciencias de la Educación (UMCE), Laboratorio de Biología, Avenida José Pedro Alessandri 774, 7750000 Ñuñoa, Chile; alexandra.guerra.ramirez@gmail.com*

Resocialization processes are used to introduce and/or rehabilitate chimpanzees, replacing the display of less appropriate behaviours with behaviours that are more suitable for social life in captivity. This is crucial for their wellbeing, because chimpanzees are social animals. The main goal of this study was to assess behavioural changes of a 35 year-old female chimpanzee after a process of resocialization with a male of the same species. The study was conducted in two stages, the first being the measurement of frequency and duration of the following behaviours regarding the female in solitary state: level of activity, feeding, handling objects, auto-grooming, self-directed behaviours, abnormal behaviours, tension, human-chimpanzee neutral, positive and negative behaviours and use of space (neutral, enriched and solitary zones); the second stage started after resocialization of both individuals, and measured individual and social behaviours (affiliative, allo-grooming, sexual, agonistic, dominance, submission) and use of space. Recording of behaviours was conducted using a video camera during one continuous hour per session, completing 123 hours of observation. The JWatcher® software was used to measure the chimpanzees' behaviours. For the analysis of results, ANOVA or Kruskal-Wallis tests were used, according to normality of data. The latency period of the first physical interaction was witnessed 2 hours from the beginning of the resocialization process in the yard between both chimpanzees. The results showed significant changes on the frequency of several behaviours of the female ($F(14,1800)=228.8$, $P<0.001$), among which are: A significant increase in behavioural diversity, displaying of social behaviours and changes in use of space, with a significant increase of use of the most enriched zone of the enclosure ($\chi^2(1, 59)=18.5$; $P<0.001$). In conclusion, this study shows that resocialization processes promoted more appropriate and positive behaviours for this captive chimpanzee.

Mounting behaviour in pigs: individual differences are not due to dominance or sexual development

Sara Hintze[1], Desiree Scott[2], Simon Turner[3], Simone Meddle[4] and Richard D'Eath[3]

[1]University of Bern, Division of Animal Welfare, Länggasstrasse 120, 3012 Bern, Switzerland, [2]The University of Edinburgh, The Royal (Dick) School of Veterinary Studies, Easter Bush Veterinary Centre, EH25 9RG Midlothian, United Kingdom, [3]SRUC (Scotland's Rural College), Animal and Veterinary Sciences Research Group, King's Buildings, West Mains Road, EH9 3JG Edinburgh, United Kingdom, [4]The Roslin Institute, Easter Bush Veterinary Centre, EH25 9RG Midlothian, United Kingdom; sara.hintze@vetsuisse.unibe.ch

Around 100 million male piglets are castrated annually in the EU, usually without anaesthesia or post-operative analgesia leading to animal welfare implications. One alternative to castration is entire male pig production. However, entire males behave differently than castrates, for example, by performing more mounting behaviour, which is suggested to be a welfare problem. 80 entire male and 80 female pigs from two batches six week apart were observed for two hours on 12 days per batch during the final six weeks before slaughter. Observations suggested that(1i) males (M) mounted around three times as much as females (F) did (total mounting mean ± s.e.: M=18.8±2.0, F=5.7±0.9) and (2) there were individual differences in mounting that were stable over time (Kendall's coefficient of concordance: W=0.33, $P<0.001$) suggesting that mounting behaviour is an individual trait rather than the appearance of random outbreaks. There were no relationships between performance of mounting behaviour and dominance rank in food competition tests (r_s=0.115, P=0.364) or the circulating levels of sex hormones (oestradiol, testosterone and progesterone, indicators of sexual development) at the end of the study. Classification of mounting into different categories revealed that sexual mounting was most common overall (51.6% of all classified mounts) and in males but only recorded once in a female. Sexual mounts lasted longer (Fisher's exact test: $P<0.001$) and provoked more high-pitched screaming by the recipient (Fisher's exact test: $P<0.001$) compared to other types of mounting (e.g. caused by crowding or during a fight) indicating a welfare problem. However, the amount of mounting received did not affect health scores (lameness: fisher's exact test: P=0.347, scratches: χ^2=0.000, P=0.983) or weight gain (M: r_s=-0.158, P=0.185, F: r_s=- 0.070, P=0.549). Overall, sexually-motivated mounting behaviour performed by certain individual entire males induces distress vocalizations in their group-mates, but the cause of this inter-individual variation is unknown.

Do horses show communicative recruitment – a case report

Rachele Malavasi[1], Alessandro Cozzi[2], Paolo Baragli[3] and Elisabetta Palagi[4,5]
[1]Consiglio Nazionale delle Ricerche, Istituto per l'Ambiente Marino Costiero, Località Sa Mardini, 09072 Torregrande (OR), Italy, [2]Irsea Research Institute in Semiochemistry and Applied Ethology, Le Rieu Neuf, 84 490 Saint Saturnin Les Apt, France, [3]University of Pisa, Equine Lab, Department of Veterinary Sciences, viale delle Piagge 2, 56124 Pisa, Italy, [4]Consiglio Nazionale delle Ricerche, Istituto di Scienze e Tecnologia della Cognizione, Via S. Martino della Battaglia 44, 00185 Roma, Italy, [5]Università di Pisa, Museo di Storia Naturale, Via Roma 79, 56011 Calci (PI), Italy; rachele.malavasi@iamc.cnr.it

In group-living species, communication is necessary to obtain a common starting time and direction of movements. Some species use communicative recruitment, where the recruiter displays a behavior that induces other group members to follow. This behavior reflects the construction of an action plan aimed at reaching a specific goal, and it has been observed so far only in capuchins and macaques. Horses are social animals, highly skilled at reading and using body language. Even though horses regulate movements through shared-consensus processes, it is unknown if an active recruitment exists. Videos collected for a different purpose on individuals living in semi-free ranging conditions, revealed 3 events that seem to reflect a process of recruitment and following, thus suggesting that a process of active communication between horses could be possible. In these events, an individual A starts walking outside the group. It stops after some meters, body and head oriented towards a 'goal' (water-pool or other individuals). While static (with pauses lasting 20 ± 9 seconds), A turns its head back one or more times (2.3 ± 2.3), giving back-glances in the direction of an individual B. Then it looks again forward, and remains still. This process of walking, pausing and back-glancing can be repeated more than once in a single event. When B starts walking towards its direction (after 6 ± 4 seconds from the last glance), A resumes walking, and the two reach the 'goal' together. The dynamic of these case reports seems to resemble the recruitment process observed in monkeys with (1) pauses; (2) back-glances; (3) maintenance of a specific position of the body during pauses; and (4) the presence of a final goal. Further data are necessary to better describe and quantify the possible communicative recruitment in this species.

Agonistic interactions and order of access to the feeder in feedlot beef calves

María V Rossner[1], Natalia M A Aguilar[2] and María B Rossner[3]
[1]Facultad de Ciencias Veterinarias. UNNE., Animal Production, Sargento Cabral 2139, 3400 Corrientes, Argentina, [2]Instituto Nacional de Tecnología Agropecuaria INTA, Marcos Brioloni S/N, 3505 Colonia Benitez Chaco, Argentina, [3]Instituto Nacional de Tecnología Agropecuaria INTA, Ruta 14 km 1086, 3313 Cerro Azul Misiones, Argentina; mvrossner@hotmail.com

The aim of the study was evaluate the relationship between access order to feeder and agonistic behaviors in feedlot beef cattle. Forty males and forty females Braford crossbreed calves 8 months of age were assigned to eight feedlot pens (10 individuals each) according to bodyweight (BW) (131.38±12.95 kg). During the first ten days in the feedlot the Order of Access to the Feeder (OAF) was recorded the first time they arrived the feeder at the time of food delivery (08:30 a.m.) scoring the order of each animal from 1st to 10th. Then continuous focal observation of agonistic interactions was recorded throughout 120 minutes. The variables registered were: Head butt (HB), Displacement (D), Chasing (CH) and Fighting (F). Then the Total Given Interactions (TG) = HB + D + CH; and received (TR) = HB +D + CH were calculated individually and summed to Agonistic Total Interactions (AGO) = Σ TG + TR + F. Correlations between the behavioral variables and BW were determined by Spearman Rank. A negative correlation Sr=-0.25 ($P \leq 0.05$) was found between OAF and TG. Also a positive correlation Sr=0.26 ($P \leq 0.05$) between TG and TR was observed. Although all calves had an average BW gain of 78 kg no correlation was found between BW and OAF. Surprisingly, those animals that arrived first to the feeder were not necessarily those who displayed more agonistic interactions and neither that had lower BW gain. This fact suggests that those animals who reached the feeder later might eat after focal observation period, and that the agonistic interactions received did not affect they performance. Both OAF and AGO are used as hierarchical indicators in cattle but according to these results the use of one or another might lead to different results.

Perception of extension advisors on farm animal welfare and relevance of the Five Freedoms

Rosangela Poletto[1] and Maria J. Hötzel[2]

[1]*Universidade Federal de Santa Catarina, Laboratório de Etologia Aplicada e Bem-Estar Animal, Rodovia Admar Gonzaga, 1346, Itacorubi, Florianópolis, SC 88034-001, Brazil,* [2]*Instituto Federal do Rio Grande do Sul, Rodovia RS 135, Km 25, Distrito Eng. Luiz Englert, Sertão, RS 99170-000, Brazil; poletto.auditoria@gmail.com*

Understanding the attitudes and perception of extension advisors relative to the welfare of the animal is of great relevance as they have a direct impact on the welfare of farm animal. Therefore, the current study aimed to evaluate the perception of these professionals, among other topics, or their opinion regarding the relevance of the Five Freedoms. A survey was sent via email three times within a period of 2 months to over 540 public and private (industry) extension advisors (veterinarians, agronomists, agricultural technicians) who mainly work with livestock production, especially swine, poultry and dairy, in the state of Santa Catarina, southern Brazil. Respondents were asked to select one option – strongly disagree (SD), disagree (Dg), neutral (Nu), agree (Ag) and strongly agree (SA), to define their beliefs on the relevance of providing animals with each of the Five Freedoms, as described by the Farm animal Welfare Council (1993). Response rate was approximately 25% (138 respondents). Most respondents strongly agreed that animals must have freedom from hunger and thirst (83% SA, 17% Ag), discomfort (79% SA, 20% Ag) and, pain, injury or disease (82% SA, 17% Ag). Freedom from fear and distress was also relevant to extension advisors (64% SA, 33% Ag, 3% Nu), but to a lesser extent than the previous freedoms. Freedom of animals to express normal behaviours was considered as the least important one, with a higher rate of neutral responses and disagreement that meeting this need is vital to animals (42% SA, 38% Ag, 13% Nu, 7% Dg). Meeting animals' behavioural needs (preventing frustration, boredom) is taken as a less significant issue than providing proper nutrition, health care and pain control to animals. This perception is likely related to lack of information and training of field professionals on applied ethology and its role on animal welfare and productivity. Additionally, as reported by most respondents in an open question, this outcome is a consequence of industry pressure to produce and intensify production systems.

The growing international opposition to intensive confinement systems using cages and crates

Sara Shields and Chetana Mirle
Humane Society International, Farm Animals, 2100 L Street, NW, Washington, DC 20037, USA;
sshields@hsi.org

Humane Society International (HSI) and its sister organization, The Humane Society of the United States (HSUS), make up one of the largest animal protection organizations in the world, with companion animal, wildlife, laboratory animal and farm animal welfare programs throughout Asia, Latin America, Africa, and North America. While we address many farm animal issues, our top priority is the intensive confinement of animals, especially battery cages for laying hens and gestation crates for pregnant sows. Battery cages, small wire enclosures that confine approximately five to ten hens, prevent hens from expressing natural perching, nesting, scratching and foraging behavior. Gestation crates, metal cages barely larger than the sow's own body, prevent nearly all movement, with the exception of standing and lying. With increasing success, we have spearheaded the effort to ban these forms of confinement in several countries including Bhutan, India, Brazil, Costa Rica and Mexico and work is beginning in Africa and Vietnam. In 2012, the country of Bhutan enacted a complete ban on battery cages. The Animal Welfare Board of India has declared that battery cage confinement is in violation of India's Prevention of Cruelty to Animals Act, with several Indian states supporting this interpretation. In Brazil and Costa Rica, a number of retailers have adopted cage-free egg procurement policies. In the United States, nine states have enacted legislation to phase-out the use of gestation crates, and three states have laws that will prohibit battery cages. Since 2007, the movement by US corporations to enact policies that shift their supply chains away from cages and crates has snowballed, with over 40 companies in 2012 pledging to move away from sow stalls. The trend is clear; consumers, businesses and civil society are embracing the issue, and support the work of ethologists showing that animals have behavioral needs.

Does milk intake or activity soon after birth predict the future growth and health of calves?

Marine Rabeyrin[1], Jeffrey Rushen[2] and Anne Marie De Passillé[2]
[1]Wagenenigen University, Wageningen, 6708 LX, the Netherlands, [2]Agriculture and Agri-Food Canada, Agassiz, BC, Canada; jeffrushen@gmail.com

Our objective was to determine whether calves' vigour in the days after birth predicts later growth and the risk of illness. Female Holstein calves (n=77) were placed for 5 d after birth in 2.2 m^2 individual pens where they were allowed 12 l/d of milk. Daily milk intakes were recorded and accelerometers attached to the calves measured time spent standing and lying down. After 5d, calves were placed in 32.8 m^2 group pens with 5-8 calves per group and fed 12 l/d of milk and *ad libitum* hay and starter from automated feeders. Calves were weighed at birth (birth weight = BW) and at 28 d of age and health checks were performed daily. Digestible energy intakes were calculated from feed intakes. During d2-d4 after birth, there were large differences between calves in the amount of milk drunk, which ranged from 5.4% BW/d (2.4 l/d) to 25.8% BW/d (10.2 l/d) with a median of 16.1% BW/d (6.8 l/d). Milk intake during d2-d4 was correlated (Spearman correlation: rs=0.48 P<0.01) with daily weight gain from birth to 28d and correlated (rs=0.30 P=0.02) with the residual weight gains to d28 (when digestible energy intakes from d5 to d28 were accounted for using multiple regression). Twelve calves were treated for respiratory or gastro-intestinal illness after 5d: 10 of these were below the median for milk intake during the first 5d (Chi square test: χ^2=5.33; P=0.02). Time spent standing during d2-d4 varied between calves but was not correlated with BW, milk intake during d2-d4, or weight gains to d28 (P>0.10). Very young calves can drink large quantities of milk, and the calves that drink the largest amounts in the first 4d of life have better long-term growth rates and are less likely to become ill. Milk intake may be a better sign of early calf vigour than activity levels.

Behavioural responses of sheep to simulated sea transport motion

Eduardo Santurtun[1], Valerie Moreau[2] and Clive Phillips[1]
[1]The University of Queensland, Centre for Animal Welfare and Ethics, School of Veterinary Sciences, Gatton Campus., 4343 Gatton, Queensland, Australia, [2]LaSalle Beauvais Polytechnic Institute, 19 rue Pierre Waguet, BP 30313, 60026 Beauvais Cedex, France; esanturtun@gmail.com

Little is known about the impact of ship motion on sheep welfare despite many sheep travelling long distances by ship. We tested the effects of Roll (sideways), Heave (vertical) and Pitch (fore-aft) motions on the behaviour of sheep. Sheep were placed in pairs in a crate positioned on a programmable platform that replicated the frequency and magnitude of typical ship movements. Treatments were applied in a replicated Latin Square design. Sheep behaviours were recorded individually by video cameras and analysed using a generalized linear model of the proportion of time spent and frequency for each behaviour. In Experiment one four sheep were exposed to one of the three motions for 30 minute periods, without food, or a control, stationary treatment. Compared with the Control, Heave reduced rumination ($P<0.001$), but Roll caused most stepping motions to correct balance ($P<0.001$). During Heave sheep spent more time with their head above ($P<0.001$) or under ($P<0.001$) their partner's head, with their back against the crate ($P=0.006$) and less time lying down ($P=0.012$). In experiment two, six sheep were fed with lucerne pellets to test whether the motions affected appetite and if so, whether an anti-emetic would be restorative. They were exposed to one of the three movements for 60 minutes periods, or a Control, with and without anti-emetic. Feed intake was not affected, but during Heave eating rate increased ($P=0.006$). Sheep spent more time steadying their head against a dividing mesh in the Heave tratment ($P=0.009$) and less time when the anti-emetic was provided ($P=0.01$), as well as less time steadying themselves against the side bars ($P=0.01$) when they received anti-emetic. Thus Heave and Roll require sheep to make regular adjustments to body posture with some benefits of anti-emetics. Heave also affected nutritional behaviour but further research is needed to understand this fully.

Two-step weaning reduces the behavioural distress response of beef cows at weaning

Maria J. Hötzel[1], Rodolfo Ungerfeld[2] and Graciela Quintans[3]
[1]Universidade Federal de Santa Catarina, Rod Admar Gonzaga 1346, Florianóplis, 88034-001, Brazil, [2]Departamento de Fisiología, Facultad de Veterinaria, Universidad de la República, Lasplaces 1550, Montevideo, 11600, Uruguay, [3]Instituto Nacional de Investigación Agropecuaria, Treinta y Tres, Uruguay, 33000, Uruguay; mjhotzel@cca.ufsc.br

We compared the behavioural response of cows to abrupt or two-step weaning with nose flaps. Thirty crossbred multiparous Angus × Heresford cows and their calves were assigned to 3 groups of 10 dyads each. On Day 0, 10 calves were fitted with nose-flaps (NF group); on Day 14, nose flaps were removed and calves were taken to a different pasture where they had no visual or olfactive contact with the dams. Dyads from a second group were abruptly and permanently separated on Day 14 (AW group). Another 10 dyads (NW group) were non-weaned controls. Behaviour of individual cows was recorded using 10 min scan sampling 38 times.d-1 from Day -3 to Day 5 (P1), and from Day 11 to Day 18 (P2). Data were analysed with mixed models for repeated measures; dyads were considered as the experimental unit. During P1 there was an interaction between group and time ($P<0.0001$) on the number of times in which cows and calves were separated by more than 5 cow's body lengths (BL). On Days 0 and 1, NF dyads were fewer times at >5BL than AW and NW dyads ($P=0.001$ for all comparisons). On Days 2, 3 and 4 NF dyads were observed fewer times at >5 BL than AW dyads ($P<0.0001$, $P=0.027$ and $P<0.0001$, respectively). During P2, AW cows were observed more times pacing than NF and NW cows ($P<0.0001$), and NF cows more times than NW cows ($P=0.036$). The day after separation (Day 15), AW cows vocalised more than NF and NW cows ($P<0.0001$), and both vocalised more times than NW cows ($P=0.008$). Two-step weaning method with nose flaps reduced the main behavioural changes in the cows that indicate distress.

Perch usage and preference of Brown and White laying hens housed in aviary systems

Pamela Eusebio-Balcazar[1], Anna Steiner[1], Mary Beck[2] and Sheila Purdum[1]
[1]University of Nebraska, Animal Science, Lincoln, NE 68583, USA, [2]Mississippi State University, Poultry Science, Starkville, MS 39759, USA; pamelaeusebio@huskers.unl.edu

The study of perch utilization is important to provide an adequate environment and enhance hen welfare. 432 1-day-old pullets, Lohmann Brown (B) and Bovan White (W) intermingled in equal numbers, were placed in 8 floor pens. At 5 wk, pullets were moved to 8 aviary units (Natura 60, Big Dutchman Inc). Each unit had three tiers and a litter area underneath and beyond the aviaries in which birds could move freely. A nest area was located in the top tier (T), and feeder troughs were placed in the middle (M) and lower tier (L). In T, there was one perch on the front (P1) and one on the back (P2). In M, there was one on the front (P3), two in the middle (P4, P5), and one on the back (P6). In L, there was one on the front (P7), one in the middle (P8), and one on the back (P9). A lower perch was provided above the litter area (P10). Number of birds on each perch was recorded every 4 hours during a day at 15, 25, and 35 wk. Data were analyzed as a complete randomized design. Averages of observations during the day were used for perch preference. For W hens, perching usage was the lowest at 8AM (20%), intermediate from 12 p.m. to 8 p.m. (36%) and the highest at 12 a.m. (87%). For B hens, the highest perch usage was at 12 a.m. (70%) compared to the rest of the day (22%) ($P<0.0001$). More B hens preferred to use P10 (6.9%), P8 (5.7%), P5 (5.4%) compared to the rest of perches (1.9%) ($P<0.0001$). In contrast, W hens preferred to use more P1 (16%) compared to the rest of perches (3.0%) ($P<0.0001$). Thus, genetic lines had different usage and position preference of perches in aviary housing systems.

Social attachment during weaning in a Beagle littermate – preliminary data

Patricia Koscinczuk, Maria Nieves Alabarcez, Romina Cainzos and Sonia Estefania Bret
Universidad Nacional del nordeste (UNNE), Facultad de Ciencias Veterinarias, Sargento Cabral
2139, 3400, Argentina; romicainzos@hotmail.com

During the sensitive period, puppies learn intra-specific communication and develop social attachment. Both, mother and littermate influence the social and behavioral development of the individuals, since a litter usually behaves like a miniature pack. The neonatal period begins with birth till 12 days and it is characterized by mother attachment. Transitional period extend from 12 to 21 days moment when reciprocal attachment occurred. Within the first 5 weeks of socialization period (3 weeks to 3 months) the mother begins the detachment phase of their litter. These are baseline data for further research to evaluate the frequency of social contacts between mother and their litter during the transition period. A Beagle dog littermate was observed (mother and six puppies). Sessions were filming during 24 hours (1°, 7° and 14° day after birth), covering the entire nest without a blind spot. Puppies and their mother were filmed simultaneously. Then, each individual was observed through PC monitor using a focal, direct and continuous sampling. For data collecting, a subsampling of 15 minutes was considered and the contact frequency between 'mother-puppy', 'puppy-mother' and 'puppy-puppy' were registered. Before each filming segment, a sampling scanning was made registering position of each puppy and the mother; presence/absence of mother inside the nest was graphically registered. With the increase of puppies age, mother presence in nest decreased (T= 14.46) and mother-puppy contact frequency decreased, the same as puppy-mother contact (T= 15.93). But, the frequency of puppy-puppy contact increased (T= 23.41; all: $P<0.0001$). During the transition period, the puppies become more active, and during the absence of mother augmented puppies' interactions, increasing social behavior within littermates. Mother detachment is a normal maternal behavior during weaning that must be considered specially when the dogs are in captivity.

Early exposure to different feed presentations affects feed sorting in dairy calves

Emily Miller-Cushon[1], Renée Bergeron[2], Ken Leslie[3], Georgia Mason[4] and Trevor Devries[1]
[1]University of Guelph Kemptville Campus, Animal and Poultry Science, 830 Prescott St., Kemptville, ON, K0G 1J0, Canada, [2]University of Guelph, Campus d'Alfred, Animal and Poultry Science, 31 St Paul St., Alfred, ON, K0B 1A0, Canada, [3]University of Guelph, Population Medicine, 50 Stone Rd. E, Guelph, ON, N1G 2W1, Canada, [4]University of Guelph, Animal and Poultry Science, 50 Stone Rd. E, Guelph, ON, N1G 2W1, Canada; tdevries@uoguelph.ca

This study examined how early exposure to different feed presentations affects development of feed sorting in dairy calves. In weeks 1-8 of life, 20 Holstein bull calves were offered concentrate and chopped grass hay (<2.5 cm) *ad libitum* at a rate of 7:3 either as a mixture (MIX), or as separate components (COM). Calves received 8 l/d of milk replacer until week 5, and incrementally reduced amounts in weeks 5-7. All calves received the MIX diet in weeks 9-11 and, subsequently, a novel total mixed ration (TMR; containing 40.5% corn silage, 22.0% haylage, 21.5% high moisture corn, and 16.0% protein supplement) in weeks 12-13. Fresh feed and orts were sampled on d 1-4 of week 8, 9, 11, 12, and 13. Feed sorting was assessed through nutrient analysis for the MIX diet and particle size analysis for the TMR. The particle separator had 3 screens (19, 8, and 1.18 mm) producing long, medium, short, and fine particle fractions. Sorting of nutrients or particle fractions was calculated as the actual intake as a percentage of predicted intake; values >100% indicate sorting for, while values <100% indicate sorting against. Data were analyzed in a repeated measures general linear mixed model. In week 8, calves fed COM consumed more non-fiber carbohydrates (1.0 vs. 0.95 kg/d, SE=0.03, P=0.04) and less neutral detergent fiber (0.43 vs. 0.53 kg/d, SE=0.02, P=0.007) than calves fed MIX, indicating greater selection of concentrate. However, when provided the MIX diet, calves previously fed COM did not sort, whereas calves previously fed MIX sorted (P<0.05) for non-fiber carbohydrates (103.2±0.6%) and against neutral detergent fiber (97.6±1.0%). Calves previously fed MIX maintained increased sorting after transition to the novel TMR (P<0.04), sorting against long particles (86.5±2.6%) and for short (101.8±0.7%) and fine (101.2±1.0%) particles. These results indicate that method of presenting feed to pre-weaned calves may have lasting effects on feed sorting.

Feeding practices and abnormal behaviors of stabled horses

Michele Cristina Vieira, Estéfane Luiz Da Silva and Denise Pereira Leme
Universidade Federal de Santa Catarina, Depto Desenvolvimento Rural e Zootecnia, Rod Admar
Gonzaga 1346, Florianópolis, 88034000, Brazil; denise.leme@ufsc.br

We studied the time used by stabled horses in the ingestion of daily food, the proportion of roughage:concentrate in their diets and the percentage of abnormal behaviors in 62 stabled horses from Santa Catarina Mounted Police, Brazil. Total daily feed was the same for all horse and was divided into 5 meals: commercial ration-5h, grass-9h, oat-12:30h, commercial ration-17h and hay-21h. The horses were observed during all meals of the day for six repetitions, and average time spent in each meal was calculated. The sum of the means of each meal resulted in the mean (± SD) of total daily time feeding of the horses. The presence of abnormal behaviors was observed between meals when the horses were in the stable without any activity. Total daily crude fiber and dry matter in the diet was analyzed. The average time (minutes) used by the horses to ingest commercial ration-5h was 18±7; grass-9h was 33±6; oat-12:30h was 16±5; commercial ration-17h was 20±7 and hay was 46±8. The total time feeding per day (133±25 minutes) corresponds to 9% of 24-hours, which is low compared to more than the 60% typically used by horses at grazing. The horses received 6.3 kg of dry matter and 2.4 kg of crude fiber a day, which is almost half of the necessary quantity. The proportion (%) roughage:concentrate in the diet was 30:70. This diet could be indicated for horses in hard work, which is not the case for all horses of this study. Percentages of horses presenting abnormal behaviors were: 61%-ingestion of wood shaving bedding; 22%-coprophagia; 17%-licking feeders and 12%-weaving. The relatively low proportion of time spent feeding, in addition to the lack of roughage, dry matter and crude fiber may have contributed to the development of abnormal behaviors observed in the horses of this study.

Integrating technology and science to bridge the gap in animal welfare knowledge transfer

Andreia De Paula Vieira[1], Fritha Langford[2], Judit Vas[3], Pericles Gomes[1], Donald Broom[4], Bjarne Braastad[3] and Adroaldo Zanella[2]

[1]*Universidade Positivo, R. Prof. Pedro Viriato Parigot de Souza 5300, Bloco Vermelho, Curitiba, PR, 81280330, Brazil,* [2]*SRUC, Roslin Institute Building, Easter Bush, Penicuik, Midlothian, EH259RG, United Kingdom,* [3]*Norwegian University of Life Sciences, P.O. Box 5003, 1432 As, Norway,* [4]*University of Cambridge, Madingley Road, CB30ES, United Kingdom; apvieirabr@yahoo.com.br*

To meet the challenge of providing easy-to-access information on animal welfare to the many stakeholders all over the world, the educational work-package of the Animal Welfare Indicators Project (AWIN), has just launched the Animal Welfare Science Hub, a website that allows for collaborative sharing of educational resources, information on animal welfare courses and other closely related subjects such as applied ethology that are being offered all over the globe, updates on the latest animal welfare news, webinars with animal welfare specialists and other science-based animal welfare information. In the Hub, course organizers are encouraged to keep their information up to date to improve its availability to students and other learners. We will present preliminary data on the amount of animal welfare courses that are being offered at university level across the globe. The resulting course repository will allow students to easily find animal welfare courses at the level that suits them most. The learning materials based on animal behaviour and welfare science that are hosted in the Hub are versatile, current, highly interactive and broad (e.g. ranging from 'identifying better farrowing systems for sows' to 'detecting facial expressions of pain in horses'). They are the result of the multidisciplinary work of animal welfare scientists, IT developers and instructional designers that together produce videos, animations, mobile applications and 3-D simulations that are portable, have good usability and good pedagogic quality. All Hub content is free and easily accessible by instructors in the classroom or online, by other interested parties worldwide. Both course information and learning materials will go through a process of external review to ensure that everything presented on the Hub is scientifically valid, educationally effective and user-friendly.

Welfare of shelter cats in Sweden – effects of husbandry practices and routines

Elin N Hirsch, Maria Andersson and Jenny Loberg
Swedish University of Agricultural Sciences, Department of Animal Environment and Health,
Box 234, 532 23 Skara, Sweden; elin.hirsch@slu.se

The popularity of the domestic cat (*Felis catus*) is increasing around the world. Consequently there is an increase in the number of cats ending up on streets and in animal shelters. Shelter life may involve several potential stressors for a cat, e.g. group housing, which may result in cats developing undesired behaviours or becoming susceptible to diseases. This can reduce the cat's welfare and decrease its adoptability. The Swedish shelter cat population has previously not been studied concerning husbandry, housing conditions or the occurrence of diseases. In this study, we aimed to provide an overview of the Swedish cat shelter population using a survey, after which we visited some (n=15) of the shelters, where 10 cats per shelter were sampled. The samples from the visits were collected during 2013. The survey was sent to all Swedish cat shelters found (n=64), focusing on husbandry practices, routines, turnover of cats and occurrence of diseases. Thirty-nine (61%) shelters answered. The mean number of cats housed per shelter were 35 (min=4, max=90). According to the survey, 82% of the shelters (n=32) practices group housing, with the most common group sizes 2-4 (46%) and 5-9 (46%) cats. All shelters stated that they had and used some form of quarantine. During the last year, the most common disease was 'cat flu' (44%) followed by 'eye problems' (eye infection and eye inflammation) (15%), however only one shelter reported cases of cat plague (feline panleukopenia virus). 12 shelters stated that they have not had any infectious diseases during the last year. To conclude, the results from the survey show a vast range in the number of cats housed at each shelter, and describe diverse routines and husbandry regimes of which some, e.g. group housing and composition of groups, may have negative consequences for the cats' welfare.

Looking back to map the future: a review of animal sentience research

Helen Proctor, Gemma Carder, Amelia Cornish and Rosangela Ribeiro
World Society for the Protection of Animals, 5th Floor, 222 Grays Inn Rd, London, WC1X 8HB,
United Kingdom; helenproctor@wspa-international.org

The science of animal sentience is rapidly expanding. Demonstrating objectively what animals are capable of is key to achieving positive change for animals. WSPA has undertaken a systematic review of the scientific literature in order to identify gaps in our knowledge, and to assess the acceptance of animal sentience within the scientific community. Two journal databases (Science-Direct and Ingenta-Connect) were searched using a peer-reviewed list of 171 keywords, comprised of three lists of human emotions, and behaviours, and terms specific to animal sentience. The list of keywords provided a broad representation of the subjective states of animals, but not a definitive list of sentience criteria. The returned data were screened for suitability, removing any review papers, and incorrect uses of the word, for example, fear must have been used in reference to the emotional state of the animal. Chi-square analyses were performed in order to address the following questions: (1) What do we know? There were significantly more articles 'assuming' the presence of the keywords than those exploring its existence c^2 (1, n=2,561)=2,497.4, $P<0.0001$. The majority of the articles reviewed were referring to negative keywords such as pain, rather than positive ones like pleasure (2,366 vs. 165 respectively). (2) Who was being studied? The majority of research was performed on vertebrates c^2 (1, n=2,552)=2,425.61, $P<0.0001$, mammals (91.89%), rodents (69.07%), and rats (51.46%). The implications of this in reference to animal sentience will be discussed. (3) Why was the research performed? As expected the majority of studies were performed for human benefit (e.g. pharmaceutical development) (c^2(2, n=2,559)=1,464.06, $P<0.0001$), as opposed to studies performed for animal welfare or ethological reasons. The results of this systematic review will help animal welfare scientists and animal protection organisations understand what is already known and where gaps exist in our understanding of animal sentience. This knowledge is essential if we are to improve the lives of animals.

Spontaneously cycling Saint Croix ewes induce cyclic activity in seasonal anestrous Suffolk ewes

Neftalí Clemente[1], Agustín Orihuela[1], Iván Flores-Pérez[1], Virginio Aguirre[1] and Rodolfo Ungerfeld[2]
[1]Universidad Autónoma del Estado de Morelos, Facultad de Ciencias Agropecuarias, Av. Universidad 1001 Colonia Cahmilpa, 62210, Mexico, [2]Universidad de la República Uruguay, Departamento de Fisiología, Facultad de Veterinaria, Lasplaces 1620, Uruguay; aorihuela@uaem.mx

In some circumstances, cyclic ewes induce reproductive activity in seasonal anestrous ewes. However, reproductive activity in temperate origin breeds is mostly induced with hormonal. However, tropical breeds are less or even not affected by season; thus, the purpose of this study was to examine the hypothesis that ewes of a tropical breed may be effective in inducing a reproductive response in ewes of temperate origin during the anestrous period. The experiment was performed in May at 19°N. Seventeen anestrous Suffolk ewes were assigned to one of two groups. While in the control group Suffolk ewes (n=9) remained isolated, in the treated group, Suffolk ewes (n=8) were in continuous contact with 20 Saint Croix ewes for 38 days. To assess ovarian activity, blood samples were collected from all ewes, by jugular venipuncture. Samples were taken at 07:30 on days -8, -2, 8, 13, 18, 23, 28, 33 and 38 (day 0=treated ewes joined with Saint Croix ewes). After day 0, more Suffolk ewes ovulated in the treated (100%) than in the control (22%) group (Chi square test; $P<0.05$). The social effect of using cyclic animals to induce estrus in non-cycling ewes was significant and might be of practical use. It was concluded that Saint Croix ewes were effective in inducing ovulation in anestrous Suffolk ewes.

Environment influence on dog behaviours in a veterinary clinical office

Romina Cainzos, Patricia Koscinczuk, Maria Victoria Rossner and Maria Nieves Alabarcez
UNNE, Facultad de Ciencias Veterinarias, Sargento Cabral 2139, 3400, Argentina;
pkoscinczuk@hotmail.com

The domestic dog is capable of adapt to novel environments modifying its behaviour according to the context. Genetic factors, previous experience and environment could affect this process. The complexity of the environment not only includes several physical events, but also human-animal interaction, perhaps one of the most important stimuli of this species. The aim of this study was to evaluate dogs response to an unknown person in two situations: (1) when the territory was known; and (2) when the territory was unknown, considering the clinical office as the territory. Eight randomly selected adult dogs of both sexes and different breed were evaluated in a clinical office. Only four individuals previously knew the place where the communication test was performed. First, the unknown person remains sitting on a chair reading a magazine during 2 minutes without making visual contact with the dog (passive phase). Second, human stands up, called the dog for its name and tried to make physical contact during 2 minutes (active phase). The sessions were filmed through the veterinary room window, with a familiar handy cam. Recorded sessions were observed through PC monitor. Focal and continuous observations of each dog were made and the following variables were registered: 'frequency of visual contact dog-human', 'latency of approach' and 'remaining close to the unknown' (seconds), considering 1m area around the chair. The T test applied showed no significant differences ($P \geq 0.05$) in behavioural responses between groups (known and unknown territory). However, when both, the territory and the person were unknown, dogs made more frequently visual contact with the unknown person (mean=6.00), than those who previously knew the territory (mean=1.75). Although the person was always unknown, when the territory was also unknown dogs started faster communication with humans.

Do lying times of dairy cattle on tie-stall farms in Canada predict prevalence of lameness?

Gemma Charlton[1], Veronique Bouffard[2,3], Jenny Gibbons[4], Elsa Vasseur[5], Derek Haley[5], Doris Pellerin[3], Anne Marie De Passillé[1] and Jeff Rushen[1]
[1]*Agriculture and Agri-Food Canada Research Centre, Agassiz, BC, VOM 1A0, Canada,* [2]*Valacta Inc, Sainte-Anne-de-Bellevue, QC, H9X 3R4, Canada,* [3]*Université Laval, QC, G1V 0A6, Canada,* [4]*DairyCo, Warwickshire, England, United Kingdom,* [5]*University of Guelph, Guelph, ON, N1G 2W1, Canada; gemmalcharlton@yahoo.co.uk*

Identifying lame cows in tie-stalls is difficult, as common gait scoring methods are not always practical. Accelerometers can be used to record the lying behaviour of individual cows automatically, and may be a useful diagnostic tool for detecting lame cows. We recorded the lying times of cows and the prevalence of lameness on 100 tie-stall farms in Eastern Canada. On each farm, the total duration of lying, lying bout frequency and mean duration of lying bouts were recorded on 40 lactating Holstein dairy cows (mean ± SD; 157.1±38.6 DIM), for four consecutive days, using accelerometers (n=3882). A numerical scoring system was used to score the cows for lameness in the tie-stall. Median lying duration was 12.4 h/d (range: 10.3 to 14.4 h/d), with 11.3 lying bouts/d (range: 8.5 to 14.8/d), and median lying bout duration of 1.2 h (range: 0.9 to 1.6 h/d). The median prevalence of lameness was 25% (range: 0% to 56%). Pearson coefficients of correlation revealed that the prevalence of lameness on a farm was negatively correlated with the median daily lying time (r=-0.25; P=0.012), the percentage of cows with a daily lying time greater than 14h/d (r=-0.26; P=0.009), the median daily frequency of lying bouts (r=-0.26; P=0.007), and the percentage of cows with a median lying bout frequency greater than 15/d (r=-0.29; P=0.004). Lameness was also positively correlated with the mean lying bout duration of the top 10% of cows (cows on each farm with the longest lying bout duration) (r=0.23; P=0.024) and the difference in lying bout duration between the top (longest lying bout duration) and bottom (shortest lying bout duration) 10% of cows (r=0.26; P=0.01). The measures of time spent lying are correlated with the prevalence of lameness in a herd. However, the correlations were all small and measures of lying time may need to be used in conjunction with other outcome measures to more accurately estimate lameness prevalence.

What is the temperament of Corriedale lambs in Uruguay – is it heritable?

Noelia Zambra[1], Diego Gimeno[2] and Elize Van Lier[3]
[1]Facultad de Agronomía, Producción Animal y Pasturas, Avda. E. Garzón 780, 12900 Montevideo, Uruguay, [2]Secretariado Uruguayo de la Lana, Rambla Baltasar Brum 3764, 11800 Montevideo, Uruguay, [3]Facultad de Agronomía, Producción Animal y Pasturas, Ruta 31, Km 21, 50000 Salto, Uruguay; noezambra@gmail.com

Temperament is the fearfulness and reactivity of an animal in response to humans and strange, novel or threatening environments. It can vary among genotypes and individuals of the same flock. Breeding animals capable of adapting to their environment is one way of improving their welfare, and this can be achieved through selection for temperament. We characterized the temperament of lambs in Uruguay and evaluated its heritability. We tested 4,777 Corriedale lambs (average age 180 days) from 8 farms (2010-2011). Temperament was characterized with the Isolation Box Test (IBT, r=0.76): isolating a sheep inside the box for 30 seconds, while an electronic 'agitation meter' attached to the box records vibrations produced by movements and high vocalitation of the lamb inside, giving arbitrary numeric values. The lower the agitation score the calmer the sheep. Heritability was estimated by Gibbs sampling using a mixed model including the contemporary group (year-farm-group), sex, mother's age and type of birth as class effects; and including lambs' age at the time of measurement, the weaning weight and the breeding value (8,382 individuals, pedigree file) as covariates. We used a chain of 500,000 iterations, discarding the first 50,000, saving each 50. For the variance's components we used 5 degrees of belief. The average IBT agitation score was 24 (observed range: 1-113), which categorizes these lambs as calm animals as compared to the Australian standard. The posterior mean of heritability (±SE) was 0.17 (±0.04). While heritability isn't very high, Uruguay uses a scheme of reference rams and a multi-trait animal model, increasing the accuracy of the estimation of EPD of the candidates for selection, which we think can genetically improve the trait. Selecting animals according to their temperament may improve their management, productivity and welfare.

Activity in free range laying hen flocks in relation to keel fracture prevalence

Gemma Richards, Lindsay Wilkins, Francesca Booth, Michael Toscano and Steve Brown
University of Bristol, Animal Behaviour and Welfare, Langford House, BS40 5DU, United
Kingdom; g.richards@bristol.ac.uk

There is a large body of evidence demonstrating a high prevalence of fractured keel bones in extensive housing systems. Today's commercial laying hen is physically heavier than its ancestor, the red jungle fowl (*Gallus gallus*). This, relative to wing size has reduced the flying ability of the modern hen. However, extensive systems often require birds to move between perches and raised levels to reach facilities. Little is known about the extent to which keel fractures may impact on bird mobility and welfare. Extensive research has previously described the main activities of free-range hens, but, very little literature exists on the general movement and 'flight activity' of birds in a free range environment. The current study investigated general movements of birds using CCTV technology and related this to their changing keel status at a flock level. Approximately 12,000 commercial laying hens were observed (two consecutive flocks of 6,000 birds) using CCTV cameras over the litter, slats, and nests. Pens were filmed for 24 hours at 25, 35, 45, 55 and 65 weeks of age. At each age (after filming) 100 birds were randomly caught from each pen, and carefully examined for keel fractures using a palpation technique. The mean percentage of birds with fractured keels at 25, 35, 45, 55, 65 weeks of age was 3.75, 35.75, 48.37, 60.37, and 65.62, respectively, across both production years. At 25-35 weeks there was a significant increase in fracture incidence as well as a reduction in flights observed; however it is extremely difficult to infer causality. A two-factor analysis of variance showed a significant main effect for age, $F_{(4,1322)}=13.51$, $P<0.01$; and for the time period, $F_{(3,1322)}=18.04$, $P<0.01$ on number of flights observed. Overall, the mean number of flights observed was highest at 55 weeks of age in both years. It may be possible that flight activity is lowest when the keel fractures are accumulating and then the number of flights increases again as the levels of fracture stabilise, however this requires further investigation.

Effects of live yeast (NCYC Sc47) supplementation on the behaviour of stabled horses

Catherine Hale[1] and Andrew Hemmings[2]
[1]LFA Celtic, Unit 3, Avondale Business Park, Mill Road, Ballyclare, BT39 9AU, Northern Ireland, United Kingdom, [2]Royal Agricultural College, School of Agriculture, Stroud Road, Cirencester, Gloucestershire, GL7 6JS, United Kingdom; catherine@lfaceltic.ie

Live yeast has been proven to exert a calmative behavioural influence on species such as captive fowl and ruminants. The positive behavioural effects of yeast have been linked to a mediatory effect exerted upon dopamine receptor populations in both the gut and the brain. This study sought to determine if similar behavioural effects occur in the domestic horse when fed live yeast (NCYC Sc47). Eight healthy horses were housed in 12×12" stables, exercised twice daily and fed a forage:concentrate diet. All horses were considered to be in light work and all were maintained at a body condition score of 5. The experiment was designed as a cross-over study consisting of two treatment periods. The animals were assigned to one of two treatments (no yeast or 10 g of yeast (NCYC Sc47)). Each treatment period consisted of three weeks adaptation, 5 days data collection and two days cross-over. An ethogram was constructed through prior observation of the animal's behaviour, and aberrant behaviours were identified. These included kicking the door, vocalisation, stereotypies, threatening behaviour, etc.. Behaviour was recorded using the ethogram over a continuous period of 60 minutes, during which time the horses were fed. The numbers of aberrant behaviours performed were recorded and the data analysed by Wilcoxon's test for matched pairs. The number of aberrant behaviours performed significantly (T=0; $P<0.001$) reduced when horses were fed live yeast. Horses receiving yeast performed an average of 47 aberrant behaviour attempts during the observation periods, compared with 75 when not supplemented. The results would indicate that feeding live yeast (NCYC Sc47) to stabled horses can reduce aberrant behaviours associated with mealtime stressors.

Geriatric dogs behavior problems: canine cognitive dysfunction (CCD) suggestive cases in Brazil

Luciana Santos De Assis, Joao Telhado Pereira and Thuany Limia
Federal Rural University of Rio de Janeiro, Veterinary Institute, BR 465, Km 7, 23.890-970, Seropédica, RJ, Brazil; lucianassis@gmail.com

CCD is a progressive neurobehavioral disorder affecting aged dogs characterized by behavioral problems related to decreased cognition. Since there is no data on the prevalence of this disease in Brazil, the difficulties in diagnosis and its effects on animal welfare and its relationship with his owner, this study aimed to survey CCD suggestive cases through a specific questionnaire applied to owners of dogs aged eight years or more, in two forms: printed in Small Animal Veterinary Hospital of UFRRJ and online throughout Brazil during 2012. The questionnaire used consists of 28 behavioral changes grouped into five categories, providing a suggestive diagnosis of four groups: normal, normal with mild impairment in cognitive function (FC), patients with moderate FC impairment (Mild CCD) and patients with severe FC impairment (Severe CCD). Other data were descriptively analyzed. Of the 375 dogs evaluated (238 females and 137 males, aged between eight and 19 years – mean age 13 years) 40% were diagnosed with CCD. However, only 25% did not have other diseases whose behavioral sings could be similar to CCD. Only three dogs (1%) had been diagnosed. As expected, the mean age increased with diagnostic groups (117, 130, 141 and 166 months, respectively). The main categories involved were: social and environmental interaction (65% and 82% related to Mild and Severe CCD, respectively), sleep-awake cycle (17% related to Mild CCD), disorientation (62% related to Severe CCD) and general activities (53% related to Severe CCD). This study concluded that CCD is an important and underdiagnosed behavioral disease, demonstrating the need for greater knowledge and improvement in diagnostic methods.

Reinventing the free stall: dairy cow preference and usage of modified stalls

Cristiane C. Abade[1,2], José A. Fregonesi[1,2], Marina A. G. Von Keyserlingk[1] and Dan M. Weary[1]
[1]University of British Columbia, Animal Welfare Program, Faculty of Land and Food Systems, 2357, Main Mall, V6T 1Z4, Vancouver, BC, Canada, [2]Universidade Estadual de Londrina, C.A.R.E. – Cuidado Animal e responsabilidade Ética, Caixa Postal: 10011, 86057-970, Londrina, Paraná, Brazil; cricabade@hotmail.com

Free stalls are designed to limit where cows lie down and stand within the stall; an unintended consequence of the restrictive design is that free stalls also reduce total time lying down and standing within the stall. The aim of this study was to test usage and preference for a less restrictive stall (solid wooden partitions between adjacent lying locations protruding 8 cm above the lying surface and extending 80 cm from brisket board). Stalls had same dimensions in both treatments (120 cm in width × 240 cm in length). Forty-eight cows were randomly assigned to groups of 6 cows each that were alternately assigned to either conventional free stalls or the new design. Groups were observed for 7d and then switched to the alternate treatment in the adjacent pen for another 7d observation. Adjacent pens were then merged for a 7d choice phase during which cows had access to both treatments. Behavior and cleanliness scores (udder and stall surface) were measured during the final 3d of each phase. During the choice phase cows spent more time lying in the conventional free stalls versus the redesigned stalls (9.4 ± 0.7 h/d vs. 4.1 ± 0.7 h/d; $P<0.014$); this result may have been due in part to familiarity with the conventional stalls. When restricted to a single option there was no difference in time spent lying down in the stall (13.2 ± 0.4 h/d vs. 12.9 ± 0.4 h/d; $P<0.42$) or time spent standing with two hooves in the bedded area (1.4 ± 0.2 h/d vs. 1.2 ± 0.2 h/d; $P<0.41$). When housed in the redesigned stalls cows had dirtier udders (score 1.6 ± 0.1 vs. 1.4 ± 0.1; $P<0.002$), likely because the stall surface had more fecal contamination (4.2 ± 0.3 vs. 0.2 ± 0.3 contaminated squares/stall; $P<0.0001$). In summary, the new design appeared to offer few benefits for cows or producers.

Diagnosis of perception: farm animals, companion animals and early childhood education

Sheilla Madruga Moreira, Laila Arruda Ribeira, Isabella Barbosa Silveira, Jerri Teixeira Zanusso, Leontino Alfredo Melo Madruga and Jennifer Veiga Mendes
Universidade Federal de Pelotas, Zootecnia, Universidade Federal de Pelotas, Campus Cpão do Leão s/n, 96001-970 Capão do Leão, RS, 354, Brazil; sheillammoreira@gmail.com

Reflections on education, knowledge and understanding of farm animals and companion animal are the first steps enabling to reflect and to educate young people about animal welfare. The objective of this study was to diagnose the previous knowledge of students with ages of 8-9 years, about animal production and companion animal. A questionnaire, with open and closed questions was applied to analyze the socio-cultural profile of 3rd grade elementary school students of a private school (Colégio Gonzaga), in Pelotas, Rio Grande do Sul, Brasil. Sixty two students regularly enrolled in 2012 answered questions regarding age, schooling, knowledge of different animal species and basic knowledge of production systems. Data were analyzed quantitatively, represented in percentages, and thereafter qualitatively, linking the research to a literature review. The majority (96.6%) of students declared knowing fish species, 91.5% rabbits, 89.8% chicken, 89.8% cattle, 83% sheep, 77.9% swine and 77% equines. Fish, rabbits and some birds were considered by 78.9% as pets, whereas 89.8% considered cattle, sheep and pigs production species. Regarding care of animals, 81.4% affirmed offering only ration and 18.6% also offer leftover food. We conclude that students know animal species destined to production; however, the distinction between pets and farm animals resulted confusing. The interviewees expressed concerns regarding the welfare of animals, demonstrating basic knowledge of their needs, especially in relation to feeding of both farm and companion animals.

Short period of confinement in a tie-stall affects dairy cows' lying behaviour and locomotion score

Daniel Enríquez-Hidalgo, Keelin O'driscoll, Dayane L. Teixeira, Maxime Lucas and Laura Boyle
Teagasc, Moorepark, Animal & Grassland Research and Innovation Centre, Fermoy, Co. Cork, Ireland; laura.boyle@teagasc.ie

Cows in experimental grazing herds are often confined for metabolic measurements. The objective of this study was to establish how transfer from pasture to tie-stalls (TS) affects lying behaviour and how a period of confinement affects locomotion score of different breeds: Jersey (J, n=16), Holstein-Friesian (HF, n=16) and J × HF (F1, n=16). Cows (187±26.5 DIM) were used over 4 replicates. Replicates were balanced for breed and stocking rate (SR) at pasture. Cows were transferred to TS (day 1) and fed freshly cut ryegrass according to their SR at pasture and breed: Jhigh=14; Jlow=17; HFhigh and F1high=16; and HFlow and F1low=20 kg DM/day. They returned to pasture 12 days later. Lying behaviour of cows in 3 replicates was video-recorded during the first 15h in the TS. Five aspects of locomotion were scored from 1 to 5 (1=Normal; 5=Anomalous) on days -4, -3, 12 and 16 (max. score=25) and averages are presented. Data were analysed using Mixed models in SAS. There were no effects of breed or SR on the lying variables ($P>0.05$). However lying time (01:30:18±00:18:58h/15 h), number of lying bouts (3.1±0.4/15 h), number of lying attempts (15.1±2.2/15 h) and number of attempts/lying bout (5.6±1.3/15 h) were outside the normal ranges for dairy cows. HF cows had lower locomotion scores (1.58±0.04) than J cows (1.84±0.04; $P<0.01$) which tended to have higher scores than F1 cows (1.70±0.04; $P=0.1$). Locomotion scores were lower on day -3 (1.52±0.04) than on day 12 (1.84±0.04) and day 16 (1.74±0.04) ($P<0.001$). Locomotion score on day 12 tended to be higher than day 16 ($P=0.06$). The behaviour data reflect disrupted patterns of lying in dairy cows transferred to TS irrespective of breed. Short term confinement in TS resulted in a deterioration in locomotory ability which persisted for at least 4 days following the cows return to pasture. The results of this experiment suggest that pasture based cows may be severely stressed by confinement and this may invalidate results obtained in metabolism studies.

Genetic association between temperament and scrotal circumference in Nellore cattle

Tiago Valente, Aline Sant'anna, Mateus Paranhos Da Costa, Fernando Baldi and Lucia Albuquerque
Faculdade de Ciências Agrárias e Veterinárias, UNESP, Zootecnia, Via de Acesso Professor Paulo Donato Castellane, 14884900, Brazil; mpcosta@fcav.unesp.br

The aim of this study was to estimate the genetic association between temperament indicator traits and scrotal circumference (SC) in yearling Nellore cattle. The temperament was assessed once (aged 18 months) using three methods: (1) movement score (MOV), scoring the animals from 1 (no movements – best temperament) to 5 (vigorous movements and jumping – worst temperament) according to their movement inside the crush; (2) flight speed (FS), recording the speed at which an animal exited from the crush (m/s), considering the fastest animal as the most reactive and had worst temperament; and (3) temperament score (TS), carried out in a corral pen, after the animal exit the crush, the measurement ranging from 1 (best) to 5 (worst temperament). Data were recorded from 7,415 animals for MOV and FS and from 23.420 for TS, and 30.515 for SC. Bayesian inference and Gibbs sampling was performed to estimate the heritability and genetic correlations, applying a linear model (for FS and SC) and threshold models (for MOV and TS). Contemporary groups (CG) for MOV, FS and TS included farm of birth, management group at birth, weaning and yearling age, date of yearling assessment and calf sex; the CG for SC included farm, year of birth and management groups at birth, weaning and yearling. The models included the direct additive genetic and residual random effects, and the fixed effects of CG, age of the dam at calving and age of animal at yearling as covariate. Heritability estimates ($P<0.05$) for MOV, FS, TS and SC were 0.12 ± 0.03, 0.28 ± 0.05, 0.15 ± 0.03 and 0.45 ± 0.02, respectively. The genetic correlation estimates were: MOV-SC= 0.07 ± 0.10, FS-SC=-0.11 ± 0.07 and TS-SC=-0.27 ± 0.08. These results indicate that there are no (MOV) or low favorable associations (FS and TS) between temperament indicator traits and SC, and the application of direct selection for SC would result in low genetic response in temperament.

Social butterflies and lone rangers: social dynamics in the sheep flock using proximity loggers

John C. Broster and Rebecca E. Doyle
Charles Sturt University and EH Graham Centre for Agricultural Innovation, Locked Bag 588, 2678 Wagga Wagga, Australia; rdoyle@csu.edu.au

The aim of this study was to use proximity logger technology to intensively study social relationships between individual sheep. Proximity loggers were attached to 48 adult merino ewes (1-9 yr old) in their home paddock (3.04 ha). Loggers recorded whenever one ewe was within approximately 4m of other ewes, a distance which is common to published studies. The duration of each interaction was also recorded. Data were recorded for 48 h continuously on three separate occasions. Proximity loggers can capture multiple interactions at the same time, and so it is possible for total daily contact times can exceed 24 h (e.g. if one sheep had 5 h of contact with each of 6 different sheep the daily contact time would be 30 h). A total of 13,396 interactions were recorded between the 48 sheep over the 6 days. Ewes averaged 40h27min of contact per day (range: 18h11min - 60h8min) with all other ewes. Most commonly, individual daily interactions lasted 30-60 min (33%), then 15-30 min (23%), 1-2 h (22%) and <15 min (15%). Longer durations were less common with 4% being 2-2.5 h daily and 3% >2.5 h. Using an ANOVA, age was determined not to be a significant influence on the number of long (>2.0 h, $P=0.498$) and short interactions (<5 min, $P=0.504$). Interestingly, sheep tended to be highly gregarious or highly unsociable, with a significant difference in the number of long and short interactions individual sheep performed (paired T-test: long interactions=13.0, short interactions=5.2, $P<0.001$). Other preliminary analyses do not indicate a preference, or avoidance, between individual pairs of sheep. To our knowledge, this is the first time individual relationships in adult sheep have been analysed in such a detailed way. Future research will be targeted at relating social interactions to temperament and dominance, detection of illness, and identifying disease spread.

Use of automatic and stationary cow brushes by dairy cows

Kosuke Miwa and Ken-ichi Takeda
Shinshu University, Graduate school of Agriculture, 8304 Minami-minowa, Kamiina, Nagano,
399-4598, Japan; 12aa122g@shinshu-u.ac.jp

A cow brush can contribute to increased natural grooming behaviour in freestall barns. There are two types of cow brush. The first is an automatic cow brush, equipped with a mechanical switch that can be activated by the cow. Automatic cow brushes may reduce frustration or stress due to boredom in intensive housing systems. The second type is the stationary cow brush, which has two separate brushes, featuring a fully moveable spring and cows rub themselves against it. The experimental objectives were to determine and compare the frequency and body areas of use by cows of the two types of cow brushes situated in freestall barns. The experiment took place in a freestall barn on a typical dairy farm of Japan. The freestall barn was 600 m^2 and contained about 80 Holstein dairy cows. The two types of cow brush were installed, one each at both ends of the barn and near the water trough. Behaviour was observed for about 5 h between morning and afternoon milking for 5 days. Individual number, frequency, and body area were recorded when cows used each brush. The automatic cow brush was used, on average, by 44% of the cows per day and the stationary cow brush was used, on average, by 29% of the cows per day. Frequency and use numbers for cows that used the automatic brush were higher than for the stationary cow brush (paired t-test: frequency; $P<0.001$, number of cows using brush; $P<0.001$). The frequency of grooming each body part was higher than expected (chi-squared test: automatic; $P<0.01$, stationary; $P<0.01$). Use of the automatic brush for the hips and hind legs was higher than for the stationary brush. The stationary cow brush was often used for the head and neck.

Authors index

A

Abade, Cristiane C.	180
Acebes, Pablo	51
Ades, César	137
Agenäs, Sigrid	140
Aguilar, Natalia M A	160
Aguirre, Virginio	173
Aiello, Katia	136
Alabarcez, Maria Nieves	167, 174
Albuquerque, Lucia	183
Alencar, Mauricio	149
Alfredo Melo Madruga, Leontino	181
Améndola, Lucía	150
Amorim Franchi, Guilherme	133
Andersen, Inger Lise	55, 81
Anderson, Claes	94
Andersson, Anna-Maria	138
Andersson, Maria	171
Aparecida Jamielniak, Josemeri	129
Arnould, Cécile	155
Arranz, Josune	48
Arruda Ribeira, Laila	181
Atkins, Norton E.	112
Averós, Xavier	48
Axel-Nilsson, Malin	58

B

Baker, Paula	47
Bak Jensen, Margit	63, 71
Baldi, Fernando	183
Baragli, Paolo	159
Barbosa Silveira, Isabella	181
Barnes, Anne	95
Basso Da Silva, Paula	84
Baxter, Emma M	100
Beausoleil, Ngaio J	53
Beck, Mary	166
Bécotte, François	113
Bell, David	73
Belson, Sue	74
Beltrami, Marcial	157
Beltrán De Heredia, Ina	48
Berckmans, Daniel	82
Bergeron, Renée	66, 101, 113, 147, 168
Berk, Jutta	102

Berri, Cécile	155
Berthiaume, Robert	147
Bertin, Aline	122, 155
Bignon, Laure	155
Bijma, Piter	70
Birch, Stephanie A.	112
Bleach, Emma C.	114
Blokhuis, Harry J	58, 135
Blumetto, Oscar	154
Boissy, Alain	94, 98
Bokkers, Eddie	71
Bolhuis, J. Elizabeth	60, 61, 62, 70, 126
Booth, Francesca	177
Borghezan Mozerle, Vitor	152, 153
Bosch, Guido	62
Bouffard, Veronique	175
Boyland, Natasha	83
Boyle, Laura	45, 67, 182
Braastad, Bjarne	170
Brand, Nirita	146
Bret, Sonia Estefania	167
Bricarello, Patrizia Ana	153
Broom, Donald M.	75, 170
Broster, John C.	184
Brown, Jennifer A	57
Brown, Steve	177
Brunberg, Emma	105
Buffington, Tony	78
Buijs, Stephanie	93
Butler, Kym	52
Butterworth, Andrew	47

C

Cainzos, Romina	167, 174
Caiozzi, Andrea	157
Caja, Gerardo	140
Cakebread, Peter	111
Calandreau, Ludovic	155
Calderón Díaz, Julia Adriana	45
Camerlink, Irene	70
Campos Maia, Alex Sandro	128, 151
Carder, Gemma	123, 172
Cardinale, Massimiliano	139
Cardoso Costa, Joao H.	65
Carere, Claudio	139

Printed in the United States
by Baker & Taylor Publisher Services